建筑立场系列丛书 No.40

苏醒的儿童空间
Awakening Kidspace

中文版
(韩语版第356期)

韩国C3出版公社 | 编

刘懋琼 王晓华 曹麟 陈玲 于风军 周一 | 译

4

- 004　Hy-Fi _ The Living
- 008　蒙彼利埃人体博物馆 _ BIG
- 012　圣安娜教堂的修复 _ Studio Galantini
- 016　耶罗岛上的圣约翰浸信会教堂 _ Beautell Arquitectos
- 020　康比教堂 _ K2S Architects
- 024　屋顶与蘑菇形结构 _ Office of Ryue Nishizawa + nendo
- 028　吹田阁 _ Office of Ryue Nishizawa

32　匀质空间之后

- 032　匀质空间之后 _ Douglas Murphy
- 036　比利亚卡约的住宅 _ Pereda Pérez Arquitectos
- 046　Entre Cielos 酒店&水疗中心 _ A4 estudio
- 054　家庭住宅 _ noname29
- 062　河畔俱乐部 _ TAO

72　苏醒的儿童空间

- 072　从空间中获取知识 _ Aldo Vanini
- 076　奥伦赛幼儿园 _ Abalo Alonso Arquitectos
- 084　Saunalahti 幼儿园 _ JKMM Architects
- 096　国王公园环境意识中心 _ Donaldson + Warn
- 106　Lucie Aubrac 学校 _ Laurens & Loustau Architectes
- 118　Ama'r 儿童文化馆 _ Dorte Mandrup Arkitekter
- 128　Lasalle Franciscanas 学校高架运动场 _ Guzmán de Yarza Blache

136　承孝相

- 136　测试建筑的力量 _ Hyungmin Pai + Seung, H-Sang
- 144　HyunAm，一座黑色的小屋
- 158　申东晔文学博物馆
- 172　平度市住宅文化馆

- 186　建筑师索引

4

- 004 Hy-Fi _ The Living
- 008 Montpellier Human Body Museum _ BIG
- 012 Saint Anna Chapel Recovery _ Studio Galantini
- 016 Saint John Baptist Chapel in El Hierro Island _ Beautell Arquitectos
- 020 Kamppi Chapel _ K2S Architects
- 024 Roof and Mushrooms _ Office of Ryue Nishizawa + nendo
- 028 Fukita Pavilion _ Office of Ryue Nishizawa

32 After Universal Space

- 032 *After Universal Space _ Douglas Murphy*
- 036 House in Villarcayo _ Pereda Pérez Arquitectos
- 046 Entre Cielos Hotel & Spa _ A4 estudio
- 054 Family House _ noname29
- 062 Riverside Clubhouse _ TAO

72 Awakening Kidspace

- 072 *Learning from Spaces _ Aldo Vanini*
- 076 Kindergarten in Orense _ Abalo Alonso Arquitectos
- 084 House of Children in Saunalahti _ JKMM Architects
- 096 Kings Park Environment Awareness Center _ Donaldson + Warn
- 106 Lucie Aubrac School _ Laurens & Loustau Architectes
- 118 Ama'r Children's Culture House _ Dorte Mandrup Arkitekter
- 128 Lasalle Franciscanas School Elevated Sports Court _ Guzmán de Yarza Blache

136 Seung, H-Sang

- 136 *Testing the Strength of Architecture _ Hyungmin Pai + Seung, H-Sang*
- 144 HyunAm, A Black Cottage
- 158 Shin DongYeop Literary Museum
- 172 Pingdu Housing Culture Center

186 Index

场所营造 Place Making

Hy-Fi_ The Living

Hy-Fi为纽约现代艺术博物馆PS1（当代艺术中心夏季音乐会）的"热身"活动创造了一次有趣且迷人的体验，同时也为制造业和设计业的未来树立了新的典范。

如果说20世纪是物理学的世纪，那么21世纪将是生物学的世纪。建筑师将生物技术与尖端的计算与工程原理相结合，创造出了新的建筑材料以及一种新的生物设计方法，为PS1（纽约当代艺术中心）设计了这座百分百可堆肥化处理的建筑。建筑师临时利用自然界的碳循环规律创造了一座来自尘土、最终会归于尘土的建筑，整个过程几乎完全无废物、无热量、无碳排放。这为我们的社会在处理建筑实体和所处环境的关系方面提供了一个崭新的视角。同时也为本土材料赋予了新的定义，并且在纽约州的农业和创新文化产业、纽约设计师和非营利性机构、以及皇后区的社区花园之间建立了直接联系。

这座圆形塔建筑由有机砖和反光砖建成。建筑师为了结合利用两种新材料的独特属性而设计了这些砖。有机砖是通过将玉米茎（除此之外别无他用）和专门开发的活根结构创新地结合到一起制造出来的（这种工艺是由一个勇于革新的新公司Ecovative研发出来的，建筑师与该公司通力合作为这项工程开发量身定制的加工程序）。反射砖是通过3M公司研发的新型采光镜膜，经过定制成型工艺而制造出来的（建筑师与3M公司合作开发出了这种材料的新用途）。反射砖被用作有机砖的培育托盘，然后它们组合构成最终结构，随后装船运回3M公司用于进一步研究。有机砖排列在建筑的底部而反射砖排列在顶部，反射砖从上引入阳光并反射到圆塔内部及塔底。

这座建筑物颠覆了砌砖建筑的承重逻辑，从而创造了一种反重力效应（建筑底部并不是粗重密实的，而是纤细多孔的）。这座设施在塔底部引入冷空气的同时从塔顶排出热空气，为人们在炎热的夏天提供一个凉爽的微气候。阳光照射在建筑物内墙，光线反射发生焦散现象而形成迷人的光影效果（比如光线照在塔底的游泳池发生的折射或是穿过一只红酒杯而呈现的光的纹理）。在纽约天际线下的玻璃塔楼和PS1建筑的砌砖结构的背景衬托下，这种结构呈现给人们的是一座看似熟悉却又是全新的建筑物。总之，这座建筑给人们带来了阴影、色彩、光线及视觉上的冲击，以及耳目为之一新、精神为之一振、充满了奇幻想象和乐观精神的前卫体验。

为了完成这项工程，The Living工作室组成立了出色的合作团队，包括Ecovative公司（纽约新公司，是此项工程零废料处理方面的专家）、3M公司（发明了采光镜膜）、先进金属涂层有限公司（负责检验这座建筑所使用的天然材料在纽约的夏季条件下是否稳固耐用）、Shabd Simon-Alexander和Audrey Louise Reynolds（两位手染印花大师为建筑使用的有机砖研发定制了颜色和涂料）、Build It Green Compost机构（位于皇后区的非营利性机构，将负责处理这座建筑安装之后的建筑材料，并将它们提供给当地社区花园）、联合制造公司、Kate Orff、SCAPE景观建筑事务所、英国奥雅纳工程顾问公司、加十国际设计机构、欧特克公司、Bruce Mau品牌设计公司、布鲁克林数字铸造厂，以及来自哥伦比亚大学（建筑师的母校）的一个研究生团队将协助工作室建造和拆解这座建筑。

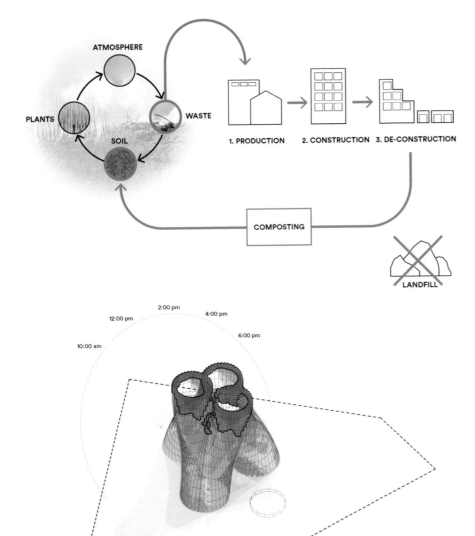

项目名称：Hy-Fi
地点：Long Island, New York, U.S
建筑师（合作团队）：
The Living/Collaborators: Ecovative, 3M,
Advanced Metal Coatings Incorporated,
Shabd Simon Alexander, Audrey Louise Reynolds,
Build It Green Compost, Associated Fabrication, Kate Orff,
SCAPE Landscape Architecture, Arup, Atelier Ten, Autodesk,
Bruce Mau Design, Brooklyn Digital Foundry,
Columbia University
竞赛时间：2014

Hy - Fi _The Living

Hy-Fi creates a fun and captivating experience for MoMA PS1 Warm Up, plus a new paradigm for the future of manufacturing and design.

If the Twentieth Century was the Century of Physics, then the Twenty-First Century is the Century of Biology. This structure uses biological technologies combined with cutting-edge computation and engineering to create new building materials and a new method of bio-design, for MoMA PS1 that is 100% grown and 100% compostable. This structure temporarily diverts the natural Carbon Cycle to produce a building that grows out of nothing but earth and returns to nothing but earth with almost no waste, no energy, and no carbon emissions. This offers a new vision for our society's approach to physical objects and the built environment. It also offers a new definition of Local Materials, and a direct relationship to New York State agriculture and innovation culture, New York City artists and non-profits, and Queens community gardens.

The structure is a circular tower of organic and reflective bricks. The architects designed these bricks to combine the unique properties of two new materials. The organic bricks are produced through a revolutionary combination of corn stalks (that otherwise have no value) and specially-developed living root structures (this process was invented by an innovative new company they are collaborating with called Ecovative, and together they

are developing a custom process for this application). The reflective bricks are produced through custom-forming of a new daylighting mirror film invented by 3M (the architects have collaborated with 3M to develop novel uses for this material). The reflective bricks are used as growing trays for the organic bricks, and then they are incorporated into the final construction before being shipped back to 3M for use in further research. The organic bricks are arranged at the bottom of the structure and the reflective bricks are arranged at the top to bounce light down on the towers and the ground.

The structure inverts the logic of load-bearing brick construction and creates a gravity-defying effect (instead of being thick and dense at the bottom, it is thin and porous at the bottom). The structure is calibrated to create a cool micro-climate in the summer by drawing in cool air at the bottom and pushing out hot air at the top. The structure creates mesmerizing light effects on its interior walls through reflected caustic patterns (like the patterns of light on the bottom of a swimming pool or shining through a wine glass). The structure offers a familiar-yet-completely-new structure in the context of the glass towers of the New York City skyline and the brick construction of the PS1 Building. And overall, the structure offers shade, color, light, views, and a future-oriented experience that is refreshing, thought-provoking, and full of wonder and optimism.

To execute this project, the architects have already built an incredible team of collaborators, including Ecovative (the New York start-up that is an expert in the no-waste material), 3M (the company that invented daylighting mirror film), Advanced Metal Coatings Incorporated (the company that is testing the natural materials for durability in New York summer conditions), Shabd Simon-Alexander and Audrey Louise Reynolds (the natural-dye artists who are developing custom colors and coatings for the organic bricks), Build It Green Compost (the Queens-based non-profit that will process the building materials after the installation and provide them to local community gardens), Associated Fabrication, Kate Orff and SCAPE Landscape Architecture, Arup, Atelier Ten, Autodesk, Bruce Mau Design, Brooklyn Digital Foundry, and a team of graduate research students at Columbia University (where the architects teach) who will help construct and deconstruct the structure.

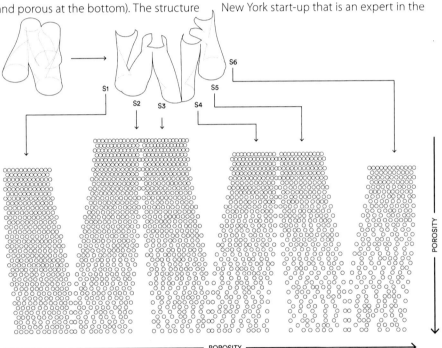

独立与统一 Islanding and Unifying

蒙彼利埃人体博物馆_BIG

蒙彼利埃人体博物馆位于蒙彼利埃市政府一带,穿插覆盖在夏帕克公园中,它可以被设想为公园与城市的融合,即自然与建筑的融合。就像油和醋这两种不兼容物质的混合物一样,城市人行道和公园草坪在彼此的拥抱中一同流动,形成一个小型梯田,俯瞰公园并提升城市之上的大自然岛屿的高度。犹如地震断层线一样,地壳被抬高并混合成为一个如同洞穴、壁龛、瞭望台和屋檐这样的潜在连续的空间。一系列看似独立的展馆就像交叉抱合的双手那样编织在一起,形成一个统一的整体。博物馆的外立面仿佛是一个弯曲的薄膜蜿蜒穿过这个场地,而不单单是划定内部与外部的周界。它在一个毫无缝隙的统一体中描绘着室内空间与室外花园,于城市与公园之间振荡摆动。博物馆的屋顶可以被设想为一个符合人体工程学的花园,花园里呈现出各种植物与矿物表面同在的动态景观,从沉思到行动,从休闲到运动,从舒缓到挑战,在这个屋顶花园中游客们可以通过各种方式来探索和表达自己的身体。博物馆的立面是透明的,最大限度地在视觉和物理空间上连接周围的城市和公园。为了避免过多强光射入建筑,建筑师计划为博物馆量身定制一个适应当地气候条件的表皮,蒙彼利埃人体博物馆弯曲的立面在东南西北各个方向波动,而百叶窗的朝向也相应地不断变化。由此产生的条纹状立面,层次弯曲,完成从水平到垂直的无缝过渡。整座博物馆的立面恰似一个适合本土气候的功能性装饰,宛若人类的指纹,看似普遍,实则独一无二。

Montpellier Human Body Museum _ BIG

Montpellier Human Body Museum is conceived as a confluence of the park and the city – nature and architecture – bookending the Charpak Park along with the Montpellier City Hall. Like the mixture of two incompatible substances – oil and vinegar – the urban pavement and the parks turf flow together in a mutual embrace forming pockets of terraces overlooking the park and elevating islands of nature above the city. Like a seismic fault line, the architectural crusts of planet earth are lifted and mingled to form an underlying continuous space of caves and niches, lookouts and overhangs. A series of seemingly singular pavilions weave together to form a unified institution – like individual fingers united together in a mutual grip. Rather than a single perimeter delineating an interior and an exterior, the facade is conceived as a sinuous membrane meandering across the site, delineating interior spaces and exterior gardens in a seamless continuum oscillating between the city and the park. The roofscape of the museum is conceived as an ergonomical garden – a dynamic landscape of vegetal and mineral surfaces that allow the park's visitors to explore and express their bodies in various ways – from contempla-

交织
interweaving

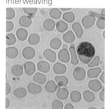
人体细胞的形成
human cell formation

人类聚集
human swarm

指纹
fingerprint

This incision interweaves the urban pavement and the parks turf, which flow together in a mutual embrace, forming terraced pockets which overlook the park and elevating islands of nature above the city.

The architectural crusts of earth are lifted and mingled to form an underlying continuous space of caves and niches, lookouts and overhangs.

Underneath the roofscape, the program is distributed in a logical and rational manner, according to the desired views towards the park, the daylight requirements, the connections to the surroundings and the internal desited connections.

A fluid space, a unifying matrix creates links between all of the programs.

The multiple entrances to the building are clearly indicated on the facade of the building.

The roofscape is conceived as an ergonomical garden – a dynamic landscape of vegetal and mineral surfaces that allows the park's visitors to explore and express their bodies in various ways.

Program
The building's program is grouped into eight major functions with the reception hall in the center.

Linear Organization
The functions are organized along a main axis, allowing the building to merge with its surroundings – creating views to the park, access to daylight, and optimizing internal connections.

From Linear Organization to Compression
The organization of the functions are compressed in order to remain within the site boundaries. For practical, functional and flexible reasons, all functions are located on one level. This compression creates connections between the functions which, if organized linearly, would not be possible.

From Compression to Organic Shapes
By multiplying the interfaces between the spaces, the shape becomes more functional, catering to the needs of the building – an adaption that results in a more fluid and organic shape, in osmosis with its environment.

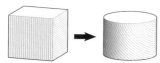

Traditional Louvers
On a cardinally oriented rectilinear building, the south facade is equipped with horizontal louvers to block the high altitude of the mid-day sun. Conversely, it is common to have vertical louvers to the east and west to block the low-incoming sunrays of morning or afternoon. If we apply this principle to a cylindrical building, we obtain a soft transition between vertical and horizontal.

Unrolled Sun Path on Cylinder
By projecting the yearly average sun angle of Montpellier on a cylinder, the direction of the louvers can be optimised. For each point on the surface of a cylinder, there is an optimal angle for the shadow dropped on the facade.

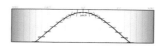

Altitude Vectors
If the sun path on a cylinder is unrolled, the red lines indicate vectors perpendicular to azimuth lines while the blue lines show the same vectors adjusted according to altitude.

Louver System
The louvers are derived by taking the closest perpendicular path through the vector field. This creates a system that can minimize direct heat gain and still maximize view. On the unrolled facade, the sun path leaves an organic-like print.

Method of Facade Generation

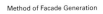 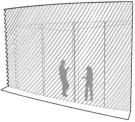

As the curvilinear geometry of the perimeter block continuously changes orientation the ideal orientation of the louvers changes along with it. Technically the louvers are cast in GFRC (glass fiber reinforced concrete). This allows them to be robust, create long spans and deal with double curvature in a simple process. The color and texture of the GFRC will be slight warm yellow because local sandstone will be added in the mixture. This underlines the local grounding of the project as well as a geological comment to the lifted landscape.

tion to the performance – from relaxing to exercising – from the soothing to the challenging. The facades of the Museum are as transparent as possible, maximizing the visual and physical interaction with the surrounding city and park. To protect from thermal exposure and glare from the abundant Montpellier sunlight, the architects propose to wrap the entire envelope in a skin tailored to the conditions of the local climate. On the sinuous facade of the Montpellier Human Body Museum that oscillates between facing North and South, East and West, the optimum louver orientation varies constantly. The resultant facade experience is a striated facade with layers that bend from horizontal to vertical in a seamless transition. Like a functional ornament adapted to its native climate the facades of Montpellier Human Body Museum resemble the patterns in a human fingerprint – both unique and universal in nature.

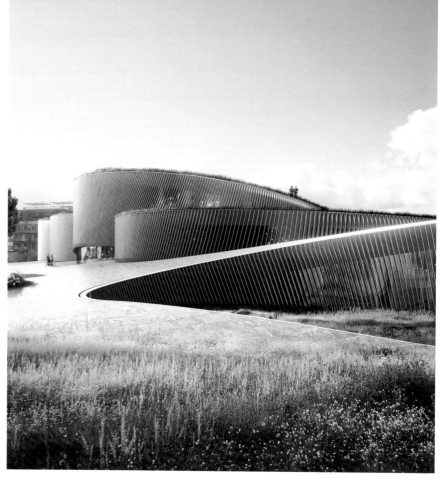

项目名称：Cité du Corps Humain
地点：Montpellier, France
建筑师：BIG
主要合伙人：Bjarke Ingels, Andreas Klok Pedersen
项目经理：Jakob Sand
项目团队：Birk Daugaard, Chris Falla, Alexandra Lukianova, Oscar Abrahamsson, Katerina Joannides, Aleksander Wadas, Marie Lançon, Danae Charatsi, Alexander Ejsing
合作建筑师：A+ Architecture, Egis Bâtiment Méditerranée, Base, L'Echo, Celsius Environnement, Cabinet Conseil Vincent Hedont
甲方：Ville de Montpellier
功能：culture　用地面积：7,800m²
竞赛时间：2013

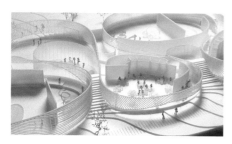

圣安娜教堂的修复 _Studio Galantini

Galantini公司位于意大利的比萨城，在结构工程师Renato Terziani的支持下，参与了圣安娜教堂的修建项目。该建筑围绕着文献学和创作之中的正确性和敏感性来建造，特别注重技术和细致的干预技术的使用。

该教堂在一个叫Lagomare的住宅小区里崛起，这个小区位于维亚雷焦镇的一个小村庄——托雷拉戈。它位于"米利亚里诺·圣罗索雷·马撒秀可里湖的自然公园"一带，临近大海。

教堂以神圣的圣安娜命名，是1973年由工程师Vardemaro Barbetta完成的项目，此命名遵循了工程师的母亲Anna的意愿。这次维修的显著特点体现在结构上：三个吊架由两个梁木加固，梁木形成带有三个铰链的门户。铰链由钢制成，放置在该结构的底部和顶部。

2010年人们发现教堂的木材结构已老化，极不安全，因此宣布禁止入内，尽管有人宣称该教堂已采取过修建行动。支撑梁木的部分由钢和焊接的缀板组成，建立起一个混合型承重结构。

修建工作于2013年四月展开，八月八日结束。教堂的修复设计没有做出任何让步，建筑技术颇为复杂，却又保留原有结构和物理材料的完整性。具有美化作用的覆盖物也被保留下来，因为它与周围的松木混为一体。完成的修建工作再现了其木质部分，并且恢复了钢铰链的机械本质。为了完成这一项目，建筑工作实现了利用特别设计的脚手架构成的悬浮结构；此外，在更换基础和退化的木质零件的过程中，脚手架还承担了负载的任务。

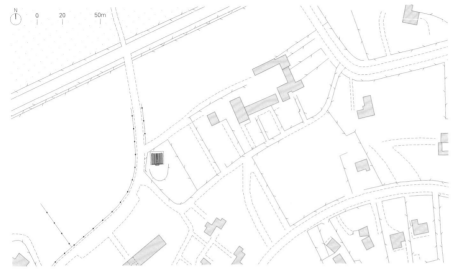

Saint Anna Chapel Recovery _ Studio Galantini

Galantini's Firm in Pisa, with the support of Renato Terziani as structural engineer, was involved in the recovery project of the Saint Anna Chapel. The work was shaped around the full philological and compositional rightness and sensitivity, paying particular attention to the usage of technology and careful intervention techniques.

The chapel arises in the Lagomare's residential complex that is located at Torre del Lago, Viareggio's hamlet. It stands inside the "Parco Naturale di Migliarino San Rossore Massaciuccoli" area, very close to the sea. The church, consecrated to Saint Anna, was built on the engineer Vardemaro Barbetta's project in 1973, following the engineer's mother's will whose name was Anna. The architectural work is highly

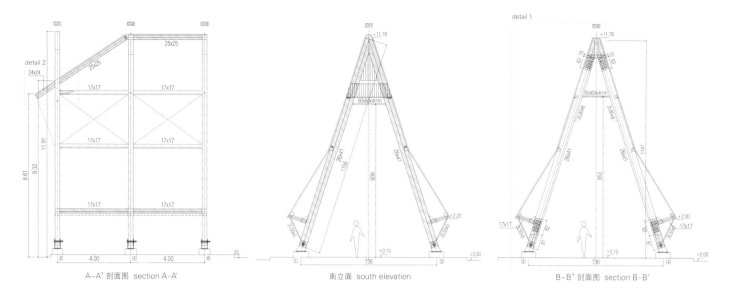

A-A' 剖面图 section A-A'　　　南立面 south elevation　　　B-B' 剖面图 section B-B'

详图1 detail 1

详图2 detail 2

characterized by the structural work: three gantries are settled by two balks that statically frame a three hinges portal. The hinges are made of steel and they are placed at the foot and at the top of the structure.

Because of the ageing of the wood in 2010 the structure was considered unsafe and declared not accessible, notwithstanding a structurally past recovery action. A support for the balks, made by steel sections and welded batten plates, set up a mixed load-bearing structure. The recovery work began in April 2013 and finished on the 8th of August. The work was designed with no compromises: the technical complexity of the work was accepted to preserve the compositional integrity of the structure and the physics of the materials. The landscaped value of the overlay was preserved too, because of its integration with the surrounding pinewood. This accomplishment was reached recreating the wooden parts and recovering the mechanical essentiality of the steel hinges. To accomplish that project, the work was realized suspending the structure using a scaffolding specifically designed for that aim; moreover the scaffolding was able to bear the efforts and the load transmitted during the substitution of the ground bases and of the degraded wooden parts.

项目名称：Saint Anna Chapel Recovery
地点：Torre del Lago, Viareggio, Lucca, Italy
建筑师：Studio Galantini
项目团队：leading architect_Paolo Galantini/ fellow architects_Marco Biondi, Mario Provenzano, Antonio Radi, Klodjan Sheshi
结构工程师：Renato Terziani
合作建筑师：Alessandro Baglieri, Carmela Bova
甲方：Parrocchia di San Giuseppe
施工：Antica Toscana
用地面积：500m²
总建筑面积：41m²
设计时间：2013
施工时间：2013.4—2013.8
摄影师：©Paolo Del Freo(courtesy of the architect)

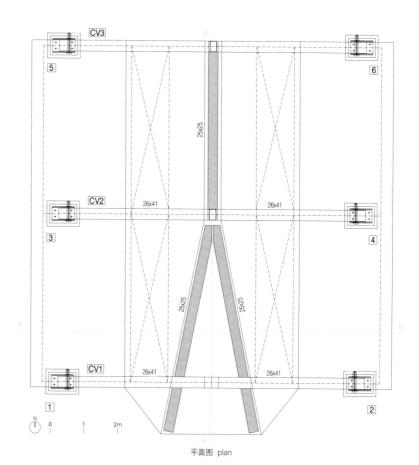

平面图 plan

教堂 Chapel

耶罗岛上的圣约翰浸信会教堂_Beautell Arquitectos

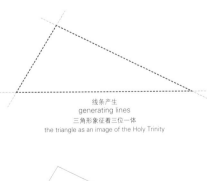

线条产生
generating lines
三角形象征着三位一体
the triangle as an image of the Holy Trinity

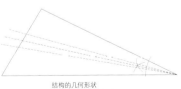

结构的几何形状
geometry of the object
神圣空间的主轴线
the main axis of the sacred space

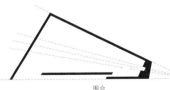

围合
enclosure
粗荔枝面的混凝土墙体/清水混凝土墙体
bush-hammered concrete walls/exposed concrete walls

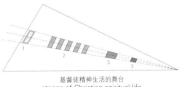

基督徒精神生活的舞台
stages of Christian spiritual life
1 洗礼区　2 社区　3 天堂的盛宴区
1. baptism 2. community 3. celestial feast

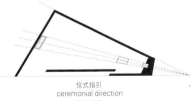

仪式指引
ceremonial direction
相关元素
related elements

固定家具
fixed furniture
由不同饰面的钢筋混凝土制成的家具构件
furniture pieces made by reinforced concrete with distinct finishings

这座可容纳40人的天主小教堂,其在平面上的投影为一座三角形的单体建筑。圣坛(源自拉丁语,altare来自altus,意为"升起")是神殿的主要组成部分,越靠近圣坛处的平面空间越狭小,但层高也相应地增加了。

三角形成为项目设计的主旨。圣坛被放在最小锐角的一侧,该角的主轴线成为神殿的中轴线,轴线将洗礼池、圣坛、神龛和十字架连成一线,如同一条生命线,象征着基督徒从洗礼、长大成人并融入社会,最后赴天国筵席等一系列生活过程。

教堂的守护神——施洗者圣约翰的生活是节制与简朴的榜样。在《圣经》中,约翰被描绘为苦行僧的形象,耶稣以他本身来反驳那些居住在皇宫圣殿、享锦衣玉食的人们。施洗者约翰敦促基督徒们都应该选择有节制的生活方式。如同约翰称自己为"沙漠中的呼声",这所教堂以他的名字命名,也将抵制曾经纵欲的生活,并成为新思潮的先导,见证一种新宗教艺术的诞生。

简朴不仅仅是一种道德信念,它已成为必然。建筑师充分利用了岛上现有的材料。没有黄金,但是有光明;没有大理石,但是有混凝土;没有精致的金银丝石膏,但是有碎火山石灰泥来装饰;没有水晶吊灯,但是当夜晚来临时,点亮灯泡也会获得同样的照明效果。

Saint John Baptist Chapel in El Hierro Island _ Beautell Arquitectos

The chapel for 40 people is projected in a plan view as a single volume in a triangular shape. Thereby the space narrows in the plan and starts to increase in the height, as people approach the altar (from Latin, altare comes from altus "rise"), which constitutes the main element of the temple.

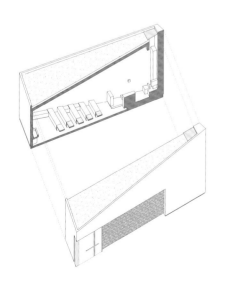

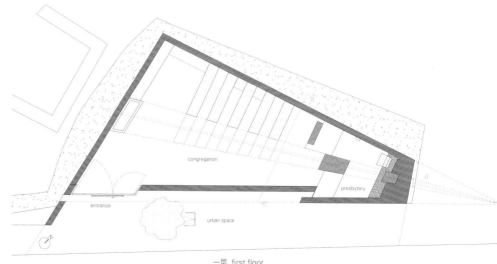

一层 first floor

The triangle becomes a leitmotiv of the project. The altar is placed in the most acute angle of the triangle and its bisector constitutes the main axis of the temple. This axis will cover the various stages of the life of a Christian, beginning with baptism, then becoming a part of the assembly and finishing in the communion of the heavenly banquet. Therefore, the baptismal font, the altar, the tabernacle and the cross are aligned, as a metaphor of the line of life.

The life of Saint John Baptist, patron saint of the chapel, was an example of sobriety and austerity. John is portrayed as an ascetic figure, Jesus countered him with those who "are in royal palaces" and "wear fine clothes." John the Baptist should urge all Christians to choose sobriety as a way of life. John defined himself as "the voice crying out in the desert", and the chapel, that will bear his name, will also protest against the excesses of the past and will be the precursor of a new stream, a testi-

mony that a new religious art is possible. Austerity was not just a moral conviction, it was a necessity. The architects use the materials they found in the island. They had no gold, but the light, they had no marble, but worth concrete, there is not plaster filigrees, but plaster of tiroliano serves them, people will not find crystal chandeliers, but, when night falls, they will light up the bulbs likewise.

项目名称：Saint John Baptist Chapel
地点：El Hierro island, Tenerife, Canary Islands, Spain
建筑师：Alejandro Beautell
项目团队：architect_Jorge Díaz/ Quantity surveyor_Eloy Fernánde
施工单位：Construcciones Expósito
甲方：Tenerife Bishopric
覆盖面积：77,15m²
造价：EUR 54,000
竣工时间：2013.6
摄影师：©Efraín Pintos (courtesy of the architect)

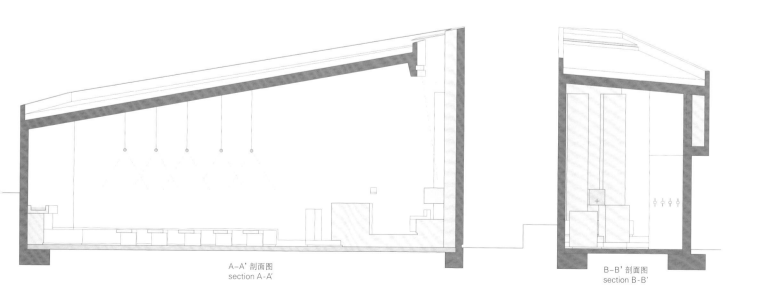

A-A' 剖面图
section A-A'

B-B' 剖面图
section B-B'

20 教堂 Chapel

康比教堂 _K2S Architects

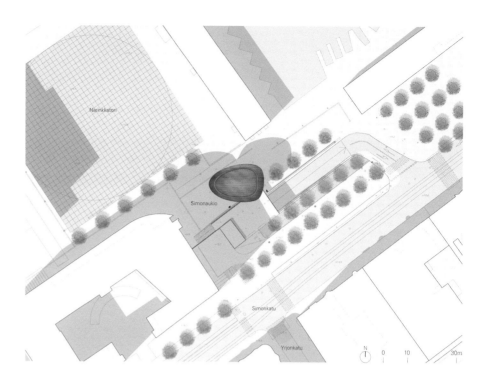

康比教堂位于赫尔辛基市中心繁华的Narinkka广场南侧。它在芬兰最热闹的城市空间里开辟出一块静默之地,让人们可以在此静心宁神。这座小型的、神圣的建筑的弧形木质立面使其自然地融入城市景观之中。与此同时,具有柔和外形的教堂内部空间环绕着游客,保护他们免受喧嚣的都市生活打扰。

教堂可以从任何方向进入。自Simonkatu街方向出发,人们来到一个面向Narinkka广场开放的小广场。在那里,一段楼梯向下通往入口层。

入口位于面对Narinkka广场和玻璃宫的两扇玻璃立面处。

只有教堂主体空间是位于木质体量内的,次要空间都位于面向广场的一处开放空间里。入口处的空间同时用作展览室,人们经常会在那里遇到牧师和社工。

圣礼室是一个静谧的地方,仿佛与喧闹的外界隔绝。轻触弧形表面,人们可以感受到构筑空间的材质所产生的温暖的感觉。教堂的内墙是由厚油桤木木板制成的。家具也是实木材质的。立面由特别锯制而成的、水平指接的云杉木板制成,并在表面着以有色透明纳米蜡。结构框架是由数控切割的胶合材料制成的。

Kamppi Chapel _ K2S Architects

The Kamppi Chapel is located on the south side of the busy Narinkka Square in central Helsinki. It offers a place to quiet down and compose oneself in one of Finland's most lively urban spaces. With its curved wood facade, the small sacral building flows into the city scape. Simultaneously the gently shaped interior space embraces visitors and shields them from the bustling city life outside.
The chapel can be approached from

项目名称：Kamppi Chapel
地点：Simonkatu 7, Helsinki, Finland
建筑师：
Kimmo Lintula,
Niko Sirola,
Mikko Summanen
设计团队：
Project architect _ Jukka Mäkinen /
Kristian Forsberg, Abel Groenewolt,
Tetsujiro Kyuma, Mikko Näveri,
Miguel Pereira, Outi Pirhonen,
Teija Tarvo, Elina Tenho, Jarno Vesa
结构工程师：Insinööritoimisto Vahanen Oy
电气和自动化设计师：
Insinööritoimisto Nurmi Oy
HVAC设计师：Insinööritoimisto Äyräväinen
音效设计：Insinööritoimisto Akukon Oy
甲方：Helsinki Parish Union and the City of H
有效楼层面积：352m²
竣工时间：2012.5
摄影师：©Tuomas Uusheimo (courtesy of the
(except as noted)

all directions. From the direction of the Simonkatu, one arrives at a small Square opening up towards the Narinkka Square. From there, a flight of stairs leads down to the entrance level.

Entrances are located in two glass facades facing the Narinkka Square and the Lasi-palatsi Building.

Only the actual chapel space is located in the wooden volume. Secondary spaces are located in a space opening up towards the square. The entrance space doubles as the exhibition space, in which one also encounters clergymen and social workers. The sacral space is a calm space, in which the lively neighborhood seems distant. Light touching down on the curved surface and the feeling of warm materials define the space. The chapel's inner walls are made of thick oiled alder planks. The furniture is also made of solid wood. The facades are made of sawn-to-order horizontal finger jointed spruce wood planks, which are treated with a pigmented transparent nanotech wax. The constructive frame consists of cnc-cut gluelam elements.

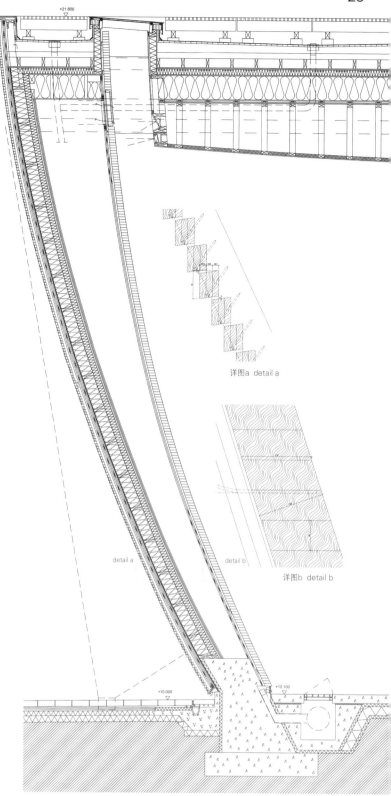

详图a detail a

详图b detail b

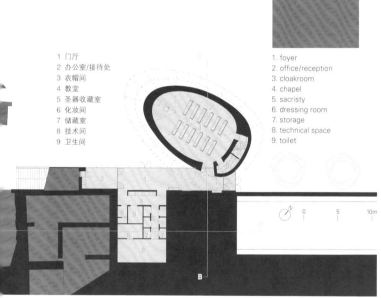

1. 门厅　　　　　1. foyer
2. 办公室/接待处　2. office/reception
3. 衣帽间　　　　3. cloakroom
4. 教堂　　　　　4. chapel
5. 圣器收藏室　　5. sacristy
6. 化妆间　　　　6. dressing room
7. 储藏室　　　　7. storage
8. 技术间　　　　8. technical space
9. 卫生间　　　　9. toilet

一层 first floor

详图1 detail 1

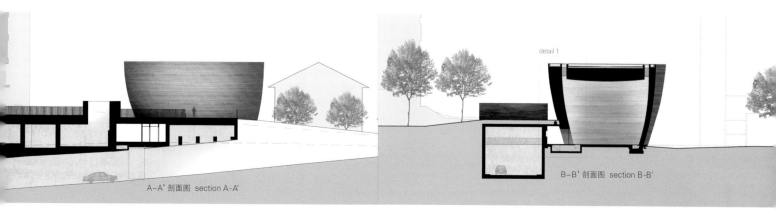

A-A' 剖面图 section A-A'

B-B' 剖面图 section B-B'

覆顶的散步道 Roofing the Walk

屋顶与蘑菇形结构 _Office of Ryue Nishizawa + nendo

这座建筑是西泽立卫办公室与nendo工作室合作为联合展览而设计的亭子。西泽立卫办公室负责该项目的建筑部分,而nendo则承担了包括家具、内景、装饰及围栏等人性化方面的工作。该建筑的主要功能是为了给学生们提供一个可以亲近自然的研究设施,同时作为处于自然环境内的一个临时的亭子,还可以供人们休憩或者放松。

该建筑位于一所大学校园后面的陡峭的山坡半山腰处。要建在自然环境内的山坡上这一特殊条件促使建筑师有了这样的想法:设计一座看上去只是一个浮动的屋顶爬上山坡的建筑。

因为建筑要依附于陡峭的山体,因此屋顶外形被弯曲成三个维度。屋顶看上去像是飘浮着的,涵盖了原有山坡的各式各样的地貌(如山体本身的平地和斜坡,原有的水泥台阶和挡土墙,以及芜杂的青草地)。由于屋顶和地面都带有坡度,因此它们之间的各点间的垂直距离是不等的,或变宽,或变窄,其间的空间类型也各不相同。为了在山坡上顺利施工,结构设计采用原木作为材料。修建这样一个依附于山体的结构是一个大型且极富挑战的项目,因此不会一次完工。

Roof and Mushrooms
_ Office of Ryue Nishizawa + nendo

This is a pavilion project being planned in conjunction with the joint exhibition by Office of Ryue Nishizawa and nendo. The Office of Ryue Nishizawa is handling the architectural aspects of the project, while nendo is in charge of the human-size aspects including furniture, interior, accessories, and railings. The structure's main function is to serve as a study facility through which students can enjoy nature, and also as a temporary pavilion in nature that provides places to rest or relax.
The site is on a hillside with a steep incline halfway up the mountain located behind the university campus. The special conditions imposed by building on a hillside in nature led to the idea for a structure in which just a roof appears to float and climb up the hill.

Because it conforms to the incline of the mountain, the roof curves in three dimensions. It seems to float, covering the various surface features of the existing hillside (like the flat grounds and slope itself, and the existing concrete steps and retaining walls, and grassy areas that have been left to run wild). With both the roof and the ground at an incline, and the distance between them narrowing and widening at different points, various types of places emerge in the space between them. The structure was designed from timber with the aim of facilitating construction on the hillside. Because it is a large and challenging project with a structure that clings to the mountain, it will not all be completed at once.

项目名称：Roof and Mushrooms
地点：Uryuyama, Kitashirakawa, Sakyo-ku, Kyoto
建筑师：Office of Ryue Nishizawa
结构工程师：ARUP
家具：nendo
安装：Yamamoto Kogyo Co.,Ltd.
装配：Ochiai Seisakusho and Kadowaki Coating
入口休息空间设计：nendo
用途：temporary pavilion
总建筑面积：57.72m²
结构：wood frame
材料：cypress
竣工时间：2013
摄影师：©Daici Ano(courtesy of the architect)

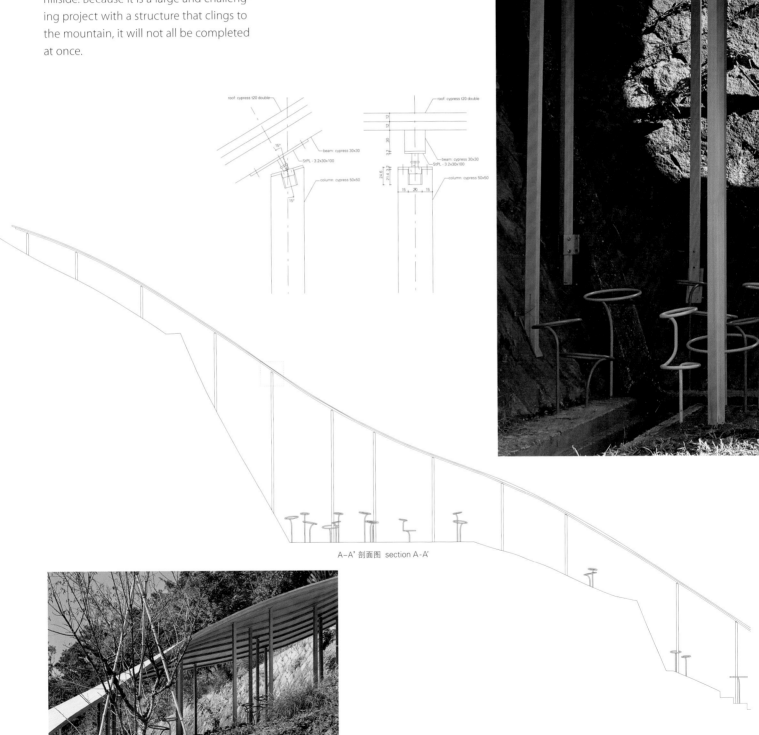

A-A' 剖面图 section A-A'

覆顶的散步道 Roofing the Walk

吹田阁_Office of Ryue Nishizawa

这座亭阁是为一座神社场地内的餐馆而建造的,毗邻小豆岛的福田小学体育馆。该建筑由两片弧形钢板叠加而成。为了形成其外形,两片钢板的四角被焊接到了一起。地上的钢板本是向上开放的,但作为屋顶的钢板却阻止了其向上开放。与此同时,屋顶钢板的中心部分随意地垂向地面,而地上的钢板刚好支撑起它的四角。两板间的空间内为游客准备了就座区,而屋顶则为孩子们提供了一个通向神殿的游乐场。这个项目虽是一个永久性的结构,但是却没有地基,这样简约的布局使亭子看起来就像是从别处直接带来并放置在这里的。

Fukita Pavilion
_ Office of Ryue Nishizawa

The project is a pavilion for a restaurant on the premise of a shrine, situated next to the gymnasium of Fukuda Elementary School in Shodoshima. The structure is comprised of two curved sheets of steel that are overlapped. To create the form, the corners of the two sheets are welded to one another. The steel plate on the floor is prevented from opening up by the plate that makes the roof while the steel plate of the roof is supported by the steel plate on the floor as it slightly slouches in the center. The space created between the two sheets provides seating for the visitors while the rooftop becomes a playground that leads to the temple for the children. It is a permanent structure but lacks foundation – it is a simple arrangement as if the pavilion was just brought and placed there.

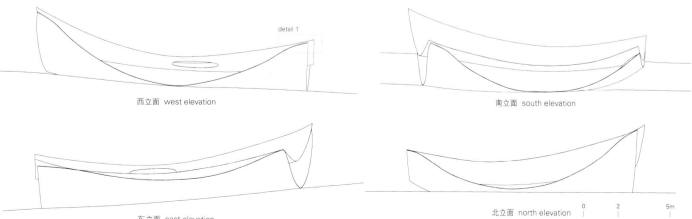

西立面 west elevation
南立面 south elevation
东立面 east elevation
北立面 north elevation

detail 1

0　2　5m

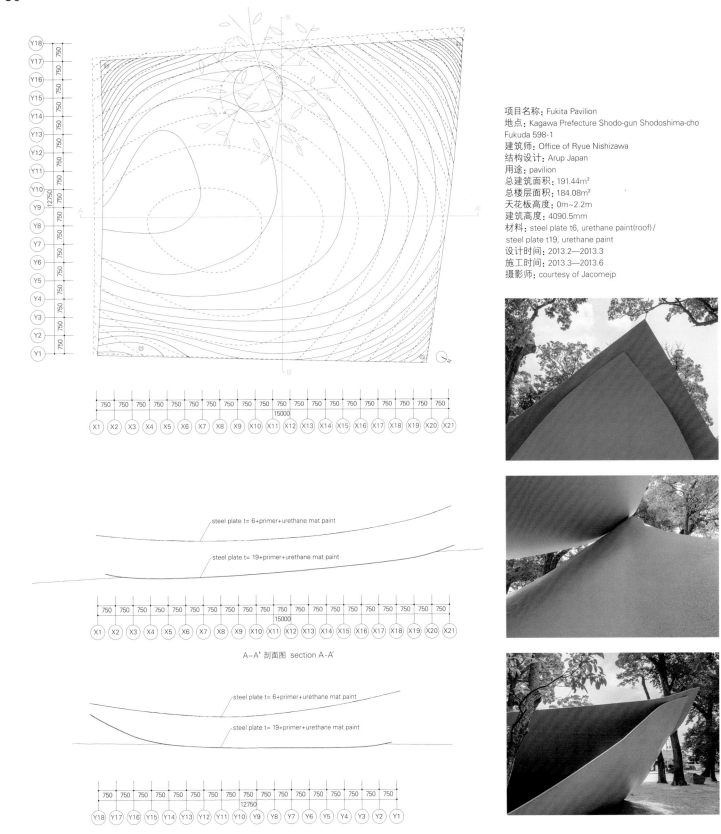

项目名称：Fukita Pavilion
地点：Kagawa Prefecture Shodo-gun Shodoshima-cho Fukuda 598-1
建筑师：Office of Ryue Nishizawa
结构设计：Arup Japan
用途：pavilion
总建筑面积：191.44m²
总楼层面积：184.08m²
天花板高度：0m~2.2m
建筑高度：4090.5mm
材料：steel plate t6, urethane paint(roof)/ steel plate t19, urethane paint
设计时间：2013.2—2013.3
施工时间：2013.3—2013.6
摄影师：courtesy of Jacomejp

A-A' 剖面图 section A-A'

B-B' 剖面图 section B-B'

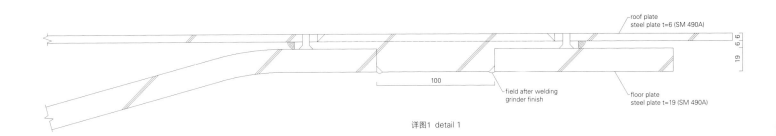

详图1 detail 1

匀质空间之后
After Univers

在全球建筑多元化的时代，方法论的界限正变得越来越不相关，正式的实验法比以前更容易获得成功，网络媒体的激增意味着通向建筑和建筑影像的道路比以前任何时候都要容易到达，20世纪对建筑师密斯·凡·德·罗的"匀质空间"的研究曾经一度被认为是建筑界简化主义的终结，但现在它们正被频繁地作为更复杂的、相互连接的建筑的基础。

曾经的现代空间承诺的一个宣言，在很多方面需要通过适应资本空间的要求才能实现，但是现在它可以被当代建筑家作为工具包来获得更多自由的和折中的建筑研究。在20世纪，"匀质空间"一直呈线性发展，其中结构和外围护结构逐渐地去物质化，直到其中一个可以识别出完全抽象的室内典范，从负荷和完全不确定性中解放出来，而现在其线性发展可以被用来创造更多的空间回旋渗透。这些"匀质空间"的新版本被几十年的形式的丰富性和奢华的简约主义所影响，更是数字化的21世纪的象征。

In a period of pluralism in global architecture, boundaries between methodologies are becoming increasingly irrelevant, formal experimentation is easier than ever to achieve, and the proliferation of online media means that access to architecture and architectural imagery is easier than at any time before, the 20th-century investigations into "universal space" of Mies van der Rohe, once thought to be the very endpoint of reductionism in architecture, are now frequently being used as a basis for more complex and articulated buildings.

What was once a statement about the spatial promises of modernity, in many respects fulfilled by the adaptation to the demands of capitalist space, can now be approached by contemporary architects as a toolkit for more free-form and eclectic tectonic investigations. The almost-linear development of "universal space" through the 20th century, where structure and envelope gradually dematerialised until one could begin to discern the ideal of a totally abstract interior, free from load and complete indeterminateness, can now be adapted to create more convoluted interpenetrations of space. Informed by intervening decades of formal exuberance and luxury minimalism, these new versions of "universal space" are more indicative of the digital 21st century.

The notion of "universal space" attributed to Mies van der Rohe was a progressive development for the old master. In various different projects throughout the twentieth century he worked on ways in which he could reduce the architectural contribution to the merest of containers: "almost nothing". From the diagrammatic plans which led to the development of the Tugendhat Villa and the Barcelona Pavilion, the envelope of the building was reduced to an almost arbitrary layer of glass, the most minimal of environmental diaphragms, while walls and junctions could be set free from their load-bearing roles. Eventually, with projects like the Farnsworth House and Crown Hall, both completed in the 1950s, Mies managed to completely abolish internal structure, leaving nothing but islands of the barest of programme inside.

In later years, "universal space" would be taken on by younger generations of architects, and it became a preoccupation of the cybernetically inspired 1960s, with their focus on flexibility, indeterminacy and adaptability. These concerns would gradually evolve and become part of the mainstream, until now when large "universal space" has become the standard approach to factory and storage facilities, most notably becoming the logic which defines office space across the world, with its constant quest for perfect wide-span grids to endlessly subdivision with desks. The logical tendency of "universal space" was satirised at the turn of the 1970s in Archizoom's No-Stop City, which took the Miesian grid off to infinity, an endpoint of capitalist urbanism where all functions were arbitrarily stretched out across one single interior space.

But there is also implied within "universal space" a potential for

比利亚卡约的住宅_House in Villarcayo/Pereda Pérez Arquitectos
Entre Cielos酒店&水疗中心_Entre Cielos Hotel & Spa/A4 estudio
家庭住宅_Family House/noname29
河畔俱乐部_Riverside Clubhouse/TAO

匀质空间之后_After Universal Space/Douglas Murphy

密斯·凡·德·罗的"匀质空间"的概念是旧概念的发展。在20世纪多种多样的项目中，他致力于减少建筑对内部的影响，直到"几乎没有"。在Tugendhat别墅和巴塞罗那展亭的图表计划中，建筑的外围护结构被减少到一个几乎随意的玻璃层，一个最小的环境隔膜，而墙体和连接处都可以从它们的负荷角色中解放出来。最终，像Farnsworth住宅和Crown大厅这样的项目，都建成于20世纪50年代，密斯成功地完全废除内部结构，只留下容纳室内最少功能的空间。

在后来的几年，"匀质空间"被年轻的一代建筑师所采用，而且它成为在20世纪60年代被网络启发的关注事物，它注重灵活性、不定性和适应性。这些逐渐地发展，成为主流的一部分，直到现在大型"匀质空间"成为工厂和仓库的标准，其中大部分性质很明显地成为定义世界办公空间的逻辑学，且持续地要求完美的大跨度网格和无边际地用桌子分割。"匀质空间"有逻辑性的倾向却在20世纪70年代（在Archizoom的No-Stop城项目中）被讽刺了，其中密斯建筑式样的方格被无限期地取消，这是资本主义城市化的终结，因为它要求所有功能都要在一处室内空间任意地延展。

但是"匀质空间"内还隐藏着一个概念差异的潜在性。逻辑性不需要直接被明显的结果所跟从，室内最低限度的项目中的自由布局能被操纵，这一前提使各种方式都变得不纯粹。本章中的四个项目——两栋西班牙房屋、一所阿根廷水疗中心以及一个中国俱乐部，都是遵循密斯·凡·德·罗的20世纪50年代的版本，尤其是遵循了Farnsworth住宅，它带有两个完全一样的、定义裸露的空间的上下层界限的平面。

A4 estudio的Entre Cielos酒店&水疗中心是以明显的内在性为标志的，建筑的主要体量有一个同等厚度的混凝土制成的屋顶和楼板，以形成一个方形的平面。它是坐落在一个由斜坡环绕的下沉庭院中，一个门廊是通过破除屋顶下面的墙体形成的，它是建筑唯一的一个洞口。室内的围护构件几乎都是实体结构，被混凝土墙环绕，通过一些纹理变化与阴影缝隙来与屋顶和地板隔开。在室内，大部分交通流线沿着外墙的边缘设置，在墙上有各种各样的小孔来射入光线，而很多水疗室都被设置成位于半开放的室内景观中的小岛空间。中央游泳池、顶部照明、暗色的混凝土墙和石质地板都印证彼得·卒姆托传奇般的瑞士瓦尔斯温泉浴场（一栋20世纪90年代的建筑佳作，也是最近记忆中对媒体最友好的建筑）所带来的影响。A4 estudio的建筑也试图获得跟这个瑞士模型一样的进深，但是建筑规模更小，所以可以想象预算也更低。

位于比利亚卡约、由Pereda Pérez建筑事务所设计的住宅也类似于向内朝向的温泉中心，但是通过在布局中创造一个有方向性的推力，来使其变得复杂。相似的地面和房顶平面在坚固的墙体的任意一端相接，创建出一个"袖管"形结构，来定义功能空间的体积。通过封锁"匀质空间"的两个边界，建筑师不需要被迫创造出一处朝外的中央功能区域。小空间——卧室、车库、厨房和浴室均被卷起，设置在"袖管"墙相对的地方，只留下一处大型开放式起居空间，其边界被嵌入建筑内，成为门廊。小巧的、脱离浴室的庭院给房间带来光线，而房间就会被"袖管"的

conceptual variations. The logic doesn't need to be followed directly to its obvious conclusion, and the premise of the free arrangement of programme within a minimally bounded interior can be manipulated, made impure in various ways. These four projects – two Spanish houses, a spa in Argentina and a clubhouse in China, all have organisational principles whose ancestors are the 1950s visions of Mies van der Rohe, and in particular the Farnsworth House, with its two almost identical planes defining the upper and lower bounds of the barest of space.

Entre Cielos Hotel & Spa by A4 estudio, is marked by a determined inwardness. The main volume of the building, defined by a concrete roof and floor slab of equal depth, defining an almost square plan, is set within a sunken courtyard approached by ramp, with a porch formed by peeling a wall back under the roof, being almost the only opening in the building. The envelope of the interior is almost completely solid, surrounded by a concrete wall, set off from the roof and floor by textural change and a slight shadow gap. Inside, much of the circulation runs along the edge of the external wall, with various perforations in the walls for bringing in atmospheric light, and many of the spa rooms are set out as island spaces within an semi-open internal landscape. The central pool, the dim lighting, the dark concrete walls and stone floor, all demonstrate the influence of Peter Zumthor's legendary Thermal Baths in Vals, a high point of 1990s architectural subtlety and one of the most media friendly buildings in recent memory. A4 estudio's building attempts similar effects of depth as the Swiss model, but within a much smaller building, and one would imagine, a smaller budget as well.

实体墙切断。在"袖管"中，墙的外覆层为深色原木，带有一系列的可以滑动、折叠和转动的隔间，以改变渗透性，由完全的开放到完全的密实。开放时，中央的家庭空间就像密斯建筑的直接理想版本，小台阶通向列柱，尝试着由一座更沉重且更大的建筑来调节纯粹的开放性。

TAO的河畔俱乐部在一条步道的设计中就采用了双平面的原则。建筑师很明确地参考Farnsworth住宅，但是其最初的透明性已经因为地点要求和功能而被毁坏。为了创建一个环形楼层平面和增加多种多样的起伏和下落，不但屋顶成为露台空间，礼堂也将在建筑内最陡峭的斜坡上形成。很明显这种折叠和环形平面具有挑战性，而且可能舍弃最初的透明性，但是这种情况也创造出其他的东西，与场地的特殊性和对普遍范例的预张力构成冲突。事实上，这是将密斯建筑的空间变形，变得更复杂。可能"匀质空间"最近几年被认为是很合理的方法，也许最成功的例子就是SANAA建筑事务所在瑞士建造的Rolex中心，一座"毯式"建筑，最初的抽象网格被刺穿并且垂直延展，形成一种更加与众不同的内部特质。

Noname29所展示的西班牙Alicante的Farnsworth式家庭住宅的设计手法运用了与中国的俱乐部相似的扭转策略，来激活结构中的不同部分。地面板的提升是考虑到了第二套居住空间的设置，使底下的空间可居住，而混凝土楼梯直通屋顶。但是形式上，尼迈耶式曲线被混合运用，双层板材在平面上呈现出倾斜的弯曲形状。场地被开发来创造一条散步道，下车的客户在返回和向上行至起居空间的地面之前，要被迫从建筑走出。在这座住宅中，起居空间和玻璃隔板位于楼层/屋顶平面（呈现出未竣工的状态）之下，栈式平面与建筑最远处的墙体连接，创造出一个折叠结构，纹理均匀，但是在地面层进行了抛光。在这个案例中，其

House in Villarcayo, by Pereda Pérez Arquitectos, is similarly inward-facing to the spa, but complicates that approach by creating a directional thrust in the organisation. A very similar ground and roof plane are met at either end with a solid wall, creating a "sleeve" which defines the volume of the functional spaces. By closing off two of the boundaries of this "universal space", the architects are not forced into a central area of programme facing outward. Smaller spaces – the bedrooms, garage, and the kitchens and bathrooms, are all tucked up against the walls of the sleeve, leaving a large open living space, with its boundaries withdrawn into the building to create a porch. A clever introduction of tiny courtyards off the bathrooms brings light to rooms that would otherwise be cut off by the solid walls of the sleeve. Within the sleeve, the walls are externally clad in darkened timber, with a series of sliding, folding and pivoting partitions changing the permeability from completely open to almost solid. When open, the central family space is like a "directed" version of the Miesian ideal, completing with the small stairs up to the podium, an attempt to reconcile pure openness with the closure of a heavier, more massive building.

Riverside Clubhouse by TAO takes this principle of doubled planes for a walk. The architects explicitly reference the Farnsworth House as the diagram, but the clarity of the original has been subjected to various deformations, both site demanded and programmatic. By creating a looped floor plane, and then adding various rises and falls, not only does the roof come into functional use as a terrace space, but an auditorium is formed where the slope within the building is at its steepest. Obviously the folding and looping challenge and perhaps discard the clarity of the original diagram, but it creates something else, in tension both with site specificity and the pretensions towards universality of the paradigm. In fact, this distortion of Miesian space into something more complex, perhaps a "universal landscape" is an approach that has been seen reasonably often in recent years, with perhaps the most successful attempt being SANAA's Rolex Centre in Switzerland, a "mat" building with an original abstract grid which is then punctured and stretched vertically, creating a far more differentiated interior quality.

The approach displayed by Noname29 to their Farnsworth-type Family House in Alicante, Spain, applies a similar twisting manoeuvre to the clubhouse in order to activate different parts of the structure. The lifting of the "ground" slab allows for a second set of living spaces to inhabit the space underneath, while a concrete staircase gives access to the roof. But formally, a dose of Niemeyer-esque curvature has entered into the mix, with the doubled slab being composed of a obliquely meandering shape in plan. The to-

巴塞罗那展亭，西班牙，密斯·凡·德·罗，1929年
Barcelona Pavilion, Spain by Mies van der Rohe, 1929

Farnsworth住宅，美国伊利诺伊州，密斯·凡·德·罗，1951年
Farnsworth House in Illinois, USA by Mies van der Rohe, 1951

功能在平面上是流畅的，创造出形状奇特的居住空间，但是它们却没有与建筑的丰富性所呼应，成为一个整体。

最近，在2013年，比利时的建筑师Robbrecht和Daem在德国为一个高尔夫球俱乐部重建了一个未建成的密斯·凡·德·罗式项目。与巴塞罗那展亭的设计时间一致，它是密斯建筑发展的最激进时期的未建成的建筑物的典范。这一重建建筑被描述为"1:1模型"，这些空间是用更现代且临时的材料（胶合木材、混凝土铺路石）来建造的。这个模型是个醒目的、能够唤起回忆的项目。但是看到那么精妙的形式逻辑采用基本的材料来表达仍然是很奇怪的。但是弄清楚密斯所提炼的极简的奢华和这些21世纪创新类型的多样性（尤其是Pereda Pérez建筑事务所利用木材和混凝土建成的房屋）之间的联系，还是十分有益的。

在公共建筑应用的密斯式极简主义被后现代主义成功消除后，虽然"匀质空间"还是很多建筑的标准，但是在20世纪90年代出现的建筑极简主义必须向其他方向发展。部分诞生于一个具有启发性的设计杂志和精装书的时代，部分受到被全球对日本及其最近建筑的兴趣的启发，且频繁地以独立的家庭房屋和小型的能唤起情感的建筑，如小教堂和水疗室为中心。但是现在很难再谈论建筑中的极简主义：这四座建筑中的每一座都展示了能被称为极简主义的元素，无论是细节和修饰的删减、光线和纹理的技巧性运用，还是通过复制楼层表面这个最容易引起争议的方式来界定内部空间。但是即使在展示这些最小的特质时，频繁的变形和丰富的外在经常使事情变得复杂。但是这仅仅是多元设计文化的一个标志。在这里，试图定义建筑实质的、持续了半个世纪的想法能够成为对简单形式进行不同的、复杂的调查的开始。

pography is exploited to create a promenade whereby the client leaving their car is forced to actually walk away from the building before they can return and ascend the slab to the living space. In this house the living space and the glass diaphragm have retreated quite far underneath the floor/roof planes, which thus take on the appearance of an unfinished building. The stacked planes are joined by a full wall at the far end of the building, creating a fold which is mostly homogeneous in texture, but polished at ground level. The island of programme in this case is equally fluid in plan, creating oddly shaped living spaces, but ones which nevertheless match the exuberance of the building as a whole.

Recently, in 2013, Belgian architects Robbrecht & Daem recreated an unbuilt Mies van der Rohe project for a golf clubhouse in Germany. Designed at the same time as the Barcelona Pavilion, it was a great unbuilt example of the most radical period of Mies' development. The recreation was described as a "1:1 model", where the spaces were created out by more contemporary, temporary materials: plywood, concrete paving stones. The model was a stunningly evocative project, but it also seemed odd to see such an exquisite formal logic expressed in rudimentary materials. But this also had the benefit of making clear the links between Mies' refined, minimalist luxury, and these 21st century variations on his typological innovations – especially the timber and concrete house by Pereda Pérez Arquitectos.

After Miesian minimalism in public buildings was effectively eliminated by postmodernism, even if "universal space" remained the principle underlying much architecture, the new minimalism in architecture that came about in the 1990s had to work from other directions. It was partly born out of a new world of aspirational design magazines and glossy books, partly inspired by a global interest in Japan and its recent architecture, and was frequently centred upon examples of single family homes and small, evocative buildings like chapels and spas. But nowadays it is difficult to even speak of minimalism in architecture: every one of these four buildings displays elements that could be described as such, whether of the reduction of detail and ornament, the focus on skilful play of light and texture, or the almost belligerent gesture of defining an interior space through the duplication of its floor surface. But even while displaying these minimal qualities, the frequent formal distortions and other exuberances often complicate matters. However this is just a sign of our pluralistic design culture, where a half-century old idea attempting to refine what is essential about architecture can become the jumping off point for various different complex investigations of simple form. Douglas Murphy

比利亚卡约的住宅
Pereda Pérez Arquitectos

记忆

这项工程的设计初衷是为了解决两个明显的现实状况：一方面，场地位于比利亚卡约郊外一处没有统一规划的移民区，地价不高，除根据法律规定的建筑高度、建筑适宜性及后退距离需要达到当下标准以外，再无其他约束。另一方面，房屋所有者是一个认同当代建筑风格的年轻家庭。他们明确了两点要求：单层住宅必然更有助于他们最大限度地亲近花园；并且此项工程的花费必须控制在非常有限的预算范围内。

因此，设计必须优先满足业主的"乐天派"和"最优化"的要求，这应该就是该设计的出发点和真实背景。与往常一样，这项工程应客户的明确要求，大体上被分为两部分：一间大起居室和直通花园的阳台为其

一；其他则作为附属的最私人的空间，主要包括一间带有浴室的主卧室，两间次卧室，另一个浴室以及一个可供孩子们学习和玩耍的空间，另外还有厨房和车库。

同时，这处小型方形场地的一侧与街道界线相接，而另一侧则是南北向的。

理念

这项工程是遵照客户的固有条件与建筑品味来设计的，它实现了一处现代化空间同时兼作家庭生活载体的构想。利用建筑与独有的室外空间的亲密关系，以及单层住宅所带来的便利性，居住者可以尽情享受这一份惬意。简而言之，该项工程力图寻求一种适合现代生活的空间组合方式，使一座住宅内的不同类型的空间满足固有生活方式的需求。

由于场地和房屋的面积有限，规划的住宅将进行扩建，来把整个场地都融入起居空间中。环绕其间的植被沿着边界生长，成为家庭生活的新界限，实现创造更大的开放式住宅的构想。

布局

调整的设计规划确保了房屋体量应该尽量紧凑地开发。该建议书呈现了非常清晰的功能布局，几乎是全面的，可以作为一个方案：最私人的空间全部被安置在靠近矩形长边两侧遮掩的混凝土墙地带，释放两者

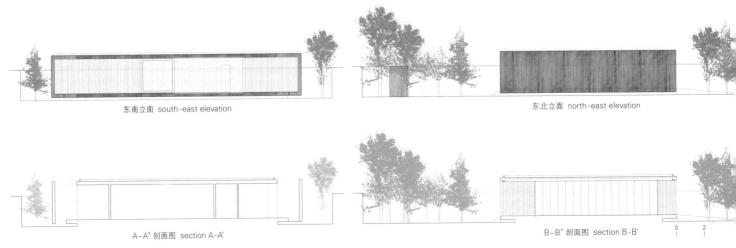

东南立面 south-east elevation　　　　　　　　　　东北立面 north-east elevation

A-A' 剖面图 section A-A'　　　　　　　　　　B-B' 剖面图 section B-B'

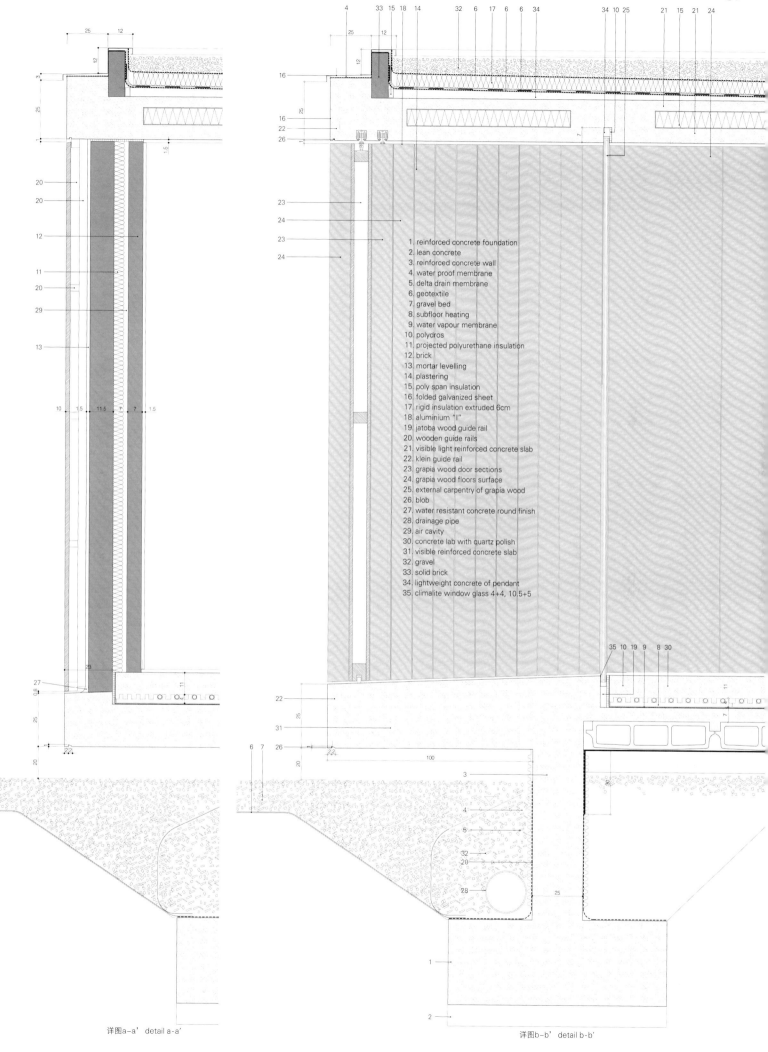

详图a-a' detail a-a'

详图b-b' detail b-b'

1. reinforced concrete foundation
2. lean concrete
3. reinforced concrete wall
4. water proof membrane
5. delta drain membrane
6. geotextile
7. gravel bed
8. subfloor heating
9. water vapour membrane
10. polydros
11. projected polyurethane insulation
12. brick
13. mortar levelling
14. plastering
15. poly span insulation
16. folded galvanized sheet
17. rigid insulation extruded 6cm
18. aluminium "I"
19. jatoba wood guide rail
20. wooden guide rails
21. visible light reinforced concrete slab
22. klein guide rail
23. grapia wood door sections
24. grapia wood floors surface
25. external carpentry of grapia wood
26. blob
27. water resistant concrete round finish
28. drainage pipe
29. air cavity
30. concrete lab with quartz polish
31. visible reinforced concrete slab
32. gravel
33. solid brick
34. lightweight concrete of pendant
35. climalite window glass 4+4, 10.5+5

之间的空间，成为一处面向花园开放的起居空间，这处空间还受到了门廊的过滤保护，贯穿东西，并且与之建立了密切的联系。

由于主要空间的集中化处理，起居空间位于屋内空间与花园的交界处，毫无疑问地成为建立亲密关系的交点。需要特别提到的是浴室：庭院横穿浴室，满足了基本的要求，并且增加了建筑整体的复杂性。

建筑材料

建筑整体所使用的混凝土和木材，其应用范围甚至延伸到了住宅内部，这种保暖却又干燥的天然舒适的材料特质提供了适宜的居住条件。同时设计师建设性地将应用于整座建筑外围的混凝土材料同时应用于室内地面和天花板，其简洁和一致性契合了这项工程追求形式简约的最初定位。中央区的家具与整体结构相结合，成为建筑的组成部分，在通往私人卧室的路上充当了隔板，同时也将起居空间的延展范围限制在私人空间之外。

最后，建筑师对采光的小心处理，制造了不同的布景透视效果，旨在不会影响到空间的纯净感。

House in Villarcayo

Memory

The starting point of the project answers to two clear situations: on one hand the plot is located in a suburban colony not yet consolidated, in the outskirts of Villarcayo, which does not have any significant value and with very few constraints but legal ones,

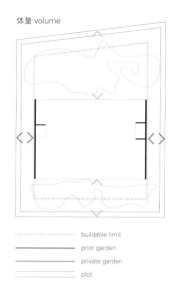
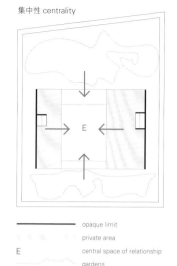
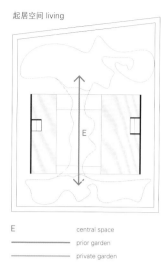

体量 volume | 集中性 centrality | 起居空间 living

- buildable limit
- prior garden
- private garden
- plot

- opaque limit
- private area
- E central space of relationship
- gardens

- E central space
- prior garden
- private garden

1 起居室 2 门廊 3 厨房 4 车库 5 主卧室 6 浴室 7 露台 8 儿童房 9 游乐室
1. living room 2. porch 3. kitchen 4. garage 5. master bedroom
6. bathroom 7. patio 8. kids' bedroom 9. playroom
一层 first floor

linked to the accomplishment of the current standards, which mainly defined the heights, suitability for building and setbacks. And on the other hand, the owners, a young family identified with the contemporary architecture, defined two questions: a house with just ground floor was a requirement in order to maximize his relationship with the garden and the project had to be unequivocally within a limited budget.

Hence, the proposal, should answer, over other premises, to the "optimism" and to the "optimization" which were demanded by the owners, supposing those as the starting point, and its "real context". The program, relatively common, was defined with accuracy by the clients, it was divided basically in two categories: on one hand, a big sitting room and terrace with direct access to the garden and on the other hand, the definition of the most private spaces, annexed, consisting in a main bedroom incorporating a bathroom, besides two bedrooms, another bathroom and a space for their children to study and play games, additionally kitchen and garage. At the same time, a single side of the small sized plot of a substantially rectangular shape touches the street line, while the other faces the south-north orientation.

Concept

The project is aligned with the conditions and the taste of the customer, it expected the resolution of a modern space, domestic container, taking advantage of the pleasure provided by the close

relationship with an own outdoor space and facilitated by the horizontal of a single floor. In short, the project seeks an adequate contemporary partnership between those spaces as something intrinsic to the way of life in a single family home versus different typologies.

The proposal, due to the limited plot and dwelling, defines a home which is expanded incorporating the whole plot to the "living space", surrounded by vegetation, which planted around the perimeter will grow until becoming the limit of the family life with the idea of making a bigger and open home.

The Organization

The volume of the house has a compact development, assuring in such manner an adjusted proposal. The proposal shows a very clear organization of the program, almost synthetic, allowing to define as a scheme: the most private spaces are accommodated close to both blind concrete sides all along its length, making free the space in between, as an open "living" to the gardens, protected by the filter of the porch and seeking the east-west orientation and the close relationship with them.

With this "central" disposition of the main space, the "living", is assured unequivocally as a meeting place to build familiar relationship, where all the spaces of the house and garden met. Special mention requires the bathrooms: they are crossed by courtyards bringing their basic needs and providing this extra of complexity to the global volume.

The Materiality

The materials which define the volume, concrete and wood, are extended to the interior qualifying the domestic conditions, providing them with a natural comfort characterized by the warm aspect but at the same time dry of the materials. Constructively, the concrete which defines the global volume, as an artifice, is incorporated in the interior defining the plane of floor and ceiling, with simplicity and coherence followed from the origin of the project being identified with the taste of the formal simplicity. The furniture of the central space, being incorporated to the architecture is transformed as part of it, works as diaphragm which guides the way to the privacy of the bedrooms and at the same time allows stretching the living space until the limits of the private.

Finally, the careful study of the lighting, defines different scenographies, being designed to not disturb the purity of the space.

项目名称：House in Villarcayo
地点：Travesía Monte Monjardín 8 Bajo 31006, Pamplona, Spain
建筑师：Carlos Pereda, Óscar Pérez Silanes
合作建筑师：Teresa Gridilla Saavedra
建造商：Construcciones Trespaderne SL + José Luis Mainz
装配工：Rodrigo Fernández Bárcena
用地面积：702.9m²
有效楼层面积：206.31m²
设计时间：2009.9
竣工时间：2012.10
摄影师：©Pedro Pegenaute

匀质空间之后 After Universal Space

Entre Cielos酒店&水疗中心
A4 Estudio

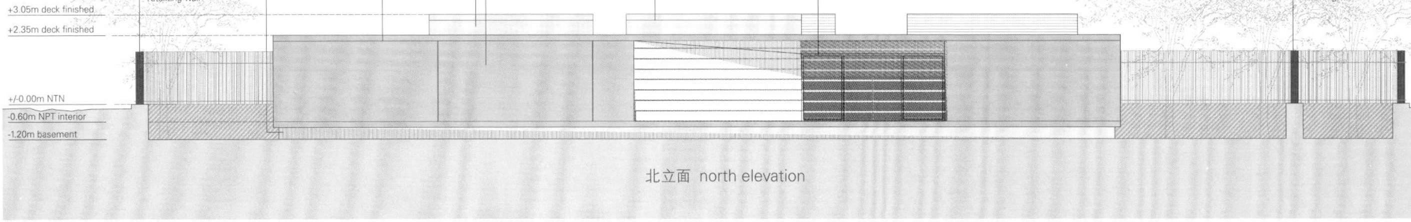

| +3.05m deck finished | reinforced concrete retaining wall | exposed concrete foundation beam | 2cm by 2cm horizontal recess | exposed concrete wall | 2cm by 2cm horizontal recess | aluminium frames | reinforced concrete retaining wall |

+3.05m deck finished
+2.35m deck finished
+/-0.00m NTN
-0.60m NPT interior
-1.20m basement

北立面 north elevation

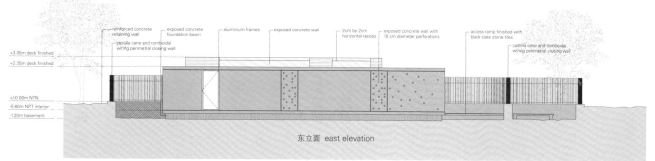

东立面 east elevation

休息和放松是此行的主要目的。建筑师会充分利用阿根廷门多萨城郊外的这个场所和这里用葡萄园、果树、白杨树和安第斯山脉的风光"编织"成的美丽的天然"挂毯"来丰富此次旅程。这个项目建在一座前面宽100m、长400m的古老的葡萄园中,一段连接建筑东西方向的人行步道形成其结构。在这种条件下,游客需要将机动车停入入口泊车处,徒步进入园内,作为一种有意脱离外界的行动。

当游客沿着这条小径漫步时,会因为遇到的一系列的户外空间而兴奋愉悦,邂逅一段有趣的体验,刺激他们的感官,并为他们在这座综合设施内的停留变化做好准备。

定义的空间及定义元素

水疗中心&酒店的设计会让人回想起儿童的积木:很多小物件可以被排列、组合,然后再重新战略性地排列,由此形成一个个特别的嵌入结构。

建筑师设想这个建筑设计能够唤起两种类型的空间:在看似随意放置在地上的这些构件之间形成的"居中"空间和内部空间。

"居中"空间是空的,它轻巧地穿梭于构件之间,或扩展或收缩,或转向或背离太阳,或敞开或关闭视野,并时时诱惑着人们继续自己的漫游和体验。就这样,人们在这些混凝土构件所界定的空间内穿行,这些混凝土构件建成了房间。这种与自然环境的关系本身已经成为一场感官的盛宴。

在酒店中,水平放置的一块混凝土板统一和界定了垂直空间,具有矛盾性。只有地板顺应了且调整了自然坡度的变化。

大众浴室+水疗中心完全与外界隔离。界定了空间的构件刻意制造出一处区域,将浴室、蒸汽房和按摩室封闭在围墙内,阻隔了与外界的视觉联系,将它变成一个灯光创造的世界。

因此,门和天花板上遍布的洞口创造了背景灯光效果和细部,为空间提供充足的照明,同时刺激人的感官。

酒店与周围的自然环境产生了一个令人沉思且具有实验性的关系,而水疗中心则令反思和冥想成为可能。游客由中层进入酒店,在一处绿色空间内沿一段坡道向下半层即可到达水疗中心。然后人们必须经过一段斜坡(斜坡由竹墙所围合),这些竹墙位于一段过渡的、半掩埋的、通往建筑的小路之前,使人们不由自主地放慢脚步,享受荫蔽。

酒店公共区域与外界之间的流动性清晰地表现了其与休闲区域之间的明显联系。然而,水疗中心的由干燥的甘蔗作物围成的墙体形成了一道不透明的面纱,光影和声音环绕在水疗中心,被隔绝在孤寂与宁静之中。

清水混凝土的标志和特色丰富了这些区域的结构色调。

一个4cm宽的松木材质的预制混凝土框架结构,其材质展现了空间的个性,而灯光增强了其色彩和纹理的分辨度。

Entre Cielos Hotel & Spa

Rest and relaxation were the main objectives of this assignment. The architects will use this location on the outskirts of the city of Mendoza and its natural tapestry of vineyards, fruit trees, poplars and views of the Andes to complement the tourism complex.

In an old vineyard, with 100 meters of frontage and 400 meters long, the project is structured from a pedestrian path that connects its buildings in an east-west orientation. This situation requires visitors to leave their vehicle in the access parking lot and enter the campus on foot, as an act of conscious separation from the outside world.

Visitors will be delighted by discovering a sequence of outdoor spaces along this path, and will encounter a playful experience that will stimulate their senses and prepare them for a transformative stay in the complex.

项目名称:Entre Cielos Tourism Development (Complejo turístico)
地点:Calle Guardia Vieja, Luján de Cuyo, Mendoza
建筑师:Leonardo Codina, Juan Manuel Filice
设计团队:Paola Meretta, Eleonora Corti, Eugenia García, José Fornés, Mariana Poppi, Gabriela Módolo, Carlos Zanonni, Luis Calvet, Mario González, Laura Benegas
结构计算:Spa-Haman_Ing. Juan Camps / Hotel_Ing. Roberto Guerrero
施工规划:A4 estudio 施工单位:Sanco SA
景观建筑师:Huemul SA
面积:Spa-Haman_597m² / Hotel_2,180m²
施工时间:2009—2011
摄影师:©García Betancourt (courtesy of the architect)

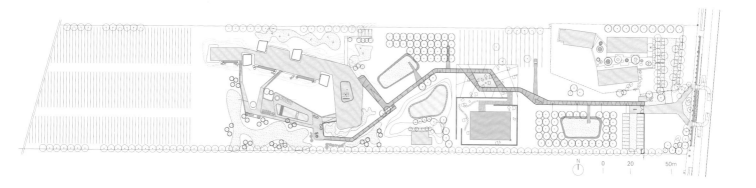

Defined Space and Defining Elements

The design of the spa and hotel is reminiscent of children's blocks: a number of small items that can be arranged, grouped, and rearranged strategically, thereby shaping the particular interventions. The architects envision that the building design evokes two types of spaces: the space defined these in-between elements positioned by seemingly arbitrary locations on the land and the interior space.

The space defined by "the in between" is empty. It slips through the elements, and is a space that expands and contracts, turning toward or away from the sun, opening or closing views and constantly tempting people to continue with the travel and experimentation. Thus the act of walking through this space is defined by concrete elements that frame the rooms. The relationship with the natural environment becomes a sensory experience in itself.

In the hotel, a horizontal concrete slab paradoxically unifies and defines the space vertically. The floor is the only element that responds and adjusts to the changes in the natural slope.

At Hamam+Spa, the space is defined in complete isolation from the outside. Elements that define the space are located with the intention of creating a course of baths, steam and massage rooms, enclosed by a perimeter wall that deprives it of a visual relationship with the outside world and turns it into a space that arises from the light.

So small openings in walls and ceilings create backlight, details, create qualifying spaces and stimulate the senses.

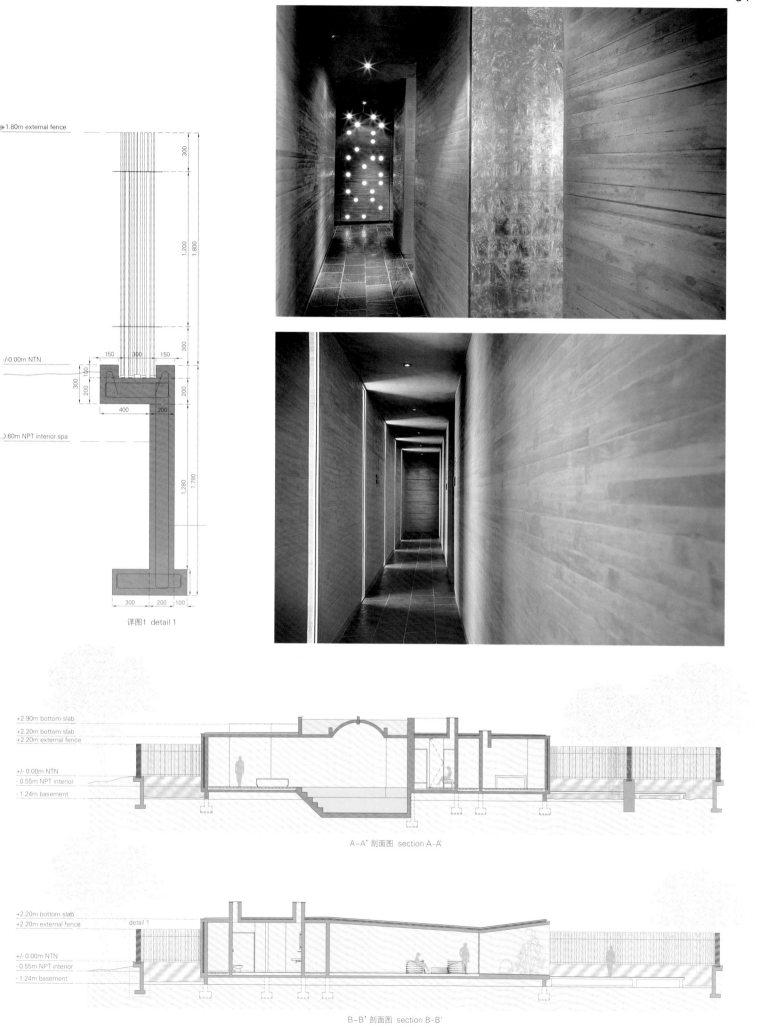

While the hotel creates a more contemplative and experimental relationship with its natural surroundings, the spa enables reflection and meditation. Visitors enter at mid-level to access the hotel via a ramp from the open green space, the spa is placed half a level down. They then have to take a ramp that is enclosed in walls of bamboo that precede an intermediate half-buried pre-access to the building that makes them slow down and enjoy the sense of shelter.

The fluidity of the hotel's public spaces with the outside articulates a clear relationship with recreational areas, however the spa's walls arising from drying cane crop create a veil of opacity, with shadows and sounds around the spa, enclosing it in a cloak of solitude and serenity.

The stamp and personality of exposed concrete provide the tectonic hue for these areas.

The light enhances the sensitivity of colors and textures provided from a precast concrete formwork in pine wood, 4 centimeters wide, built to convey the personality of the spaces.

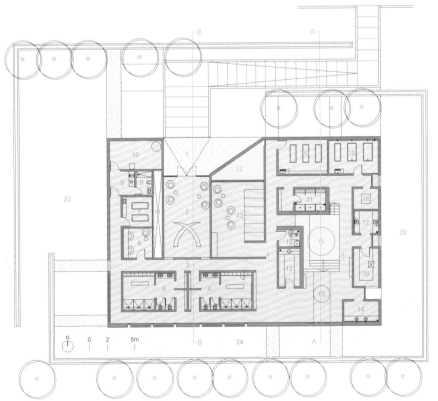

一层 ground floor

1. access
2. reception
3. walk
4. bathroom/dressing room for women
5. bathroom/dressing room for men
6. beauty salon
7. massage room
8. kitchenette
9. bathroom
10. machine room
11. light yard
12. steam room 40°C
13. hot stone
14. skin care 1
15. pool
16. steam room 60°C
17. skin care 2
18. rain forest
19. foam massage room
20. oil massage room
21. mud bath
22. reflection room
23. tearoom

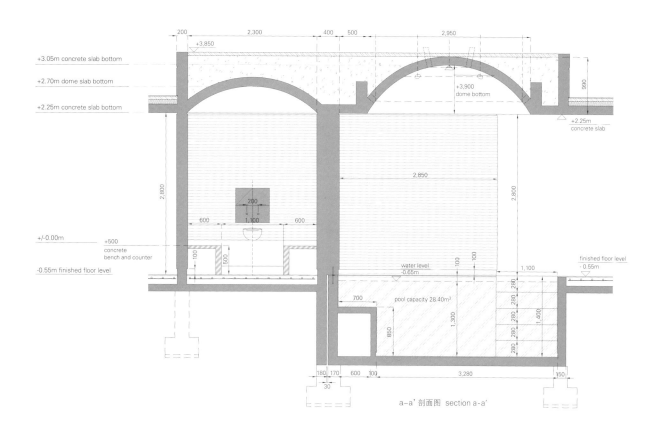

a-a' 剖面图 section a-a'

家庭住宅
noname29

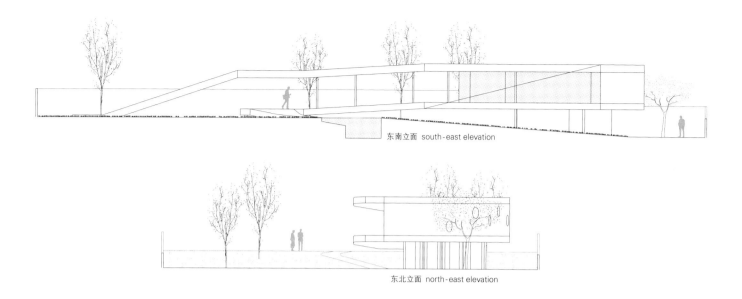

东南立面 south-east elevation

东北立面 north-east elevation

这座展馆式住宅坐落于平坦的场地上。建筑师所面临的挑战是创造一处代表其自身场地特征的、适合居住的空间,并且在某种程度上,它周围环绕的景观的特点是旱地、连绵起伏的丘陵和低灌丛。这些特点与建筑物的柔美曲线相结合,寻求建筑与自然环境之间的和谐。

力求融合内外之间、沉思性和家庭性之间、人工和天然之间的界限;这些一直是这座独特建筑的项目战略,比起住宅,这里更适合称之为展馆。

新建筑矗立在以前的两处场地上,场地原来坐落着一座没有什么价值的老房子,带有一个游泳池和一个花园,或甜菜园。这个花园是与以前的建筑相关的重要部分,对开发商来说具有高情感价值。建筑的开放式性质使室内设计毫无限制,拥有一个展厅、一间办公室、一家餐厅等。

混凝土挡板从最低点到屋顶贯穿整个地面,连接着有生命的和无生命的物体。这是一个位于景观,而不是位于木板上的挡板,呈立体式,吸引人进入每个场景,布局内部空间,空间像无限个不同的、环绕的曲线,可以根据人在居所内的活动来改变它们的形式。

新建筑的设计力求保持每一个构成花园的元素。因此,在游泳池周围出现的曲形外围护结构与场地的树木和灌木连在一起。

房屋和花园的设计逻辑被打破,以形成一个整体。尽管建筑物有着醒目的外观,整座建筑依旧融合了宜居的结构和自然的元素。每平方米的场地都是相连的,这样一来,外部空间似乎与每一处内部空间都相得益彰。每座房屋都有各自的场地的理念已经消失,导致所有建筑都面朝远处的风景、海岸线和更广阔的大海开放。建筑外围护结构的墙体在北面是封闭的,与临近建筑物隔离开来,但是却面向场地和风景开放,且有着第二层墙体,墙体标志着一个围绕房子内部且连接不同空间的微妙行程。在不知不觉中,人们跨越一个流动和连续的圆形轨道(能够成功地将人们引入不同区域),从一个空间被引导到另一个空间。房中的家具围绕着墙体来进行调整、配置、依附,旨在成为带来住户记忆的重要元素。

地板、墙壁和天花板都采用混凝土来建造,带有不同的饰面,且布局、处理方式和种类都不同:弯曲的、平坦的、水平的、垂直的、穿孔的、平滑的、表面抛光的、去活化的、打磨施工的,以弥补其在特定项目的开发中所采用的混凝土技术和实践。在施工中使用材料的丰富性使这座房屋既简单又复杂。该建筑追随了这一理念。只有某些玻璃窗格能暂时把它变成一个"胶囊",那就是当它呈现出宇宙特质的时候。

Family House

This dwelling-pavilion is located on a geographically smooth plot. The challenge was to create a habitable space which identified itself with the topography of its plot as well as, in a way, the landscape surrounding it, characterised by aridity, rolling hills and low thickets. This connects with the soft curves of the structures, in keeping with the search for harmony between the building and its natural surroundings.

To dissolve the limits between inside and outside, between the contemplative and the domestic, between the artificial and the wild: such have been the project strategy of this unique building which could be regarded as more of a pavilion than an abode.

The new construction stands on two plots previously occupied by an old house of little value, a swimming pool and a garden – orchard. This garden was an important element in relation with the previous building, and has a high sentimental value for the developer. Its open-plan nature leaves the function of the interior open-ended: an exhibition hall, an office, a restaurant, etc..

1 起居室　2 餐厅　3 厨房　4 更衣室　5 浴室　6 卧室
1. living room 2. dining room 3. kitchen 4. dressing room 5. bathroom 6. bedroom
二层　second floor

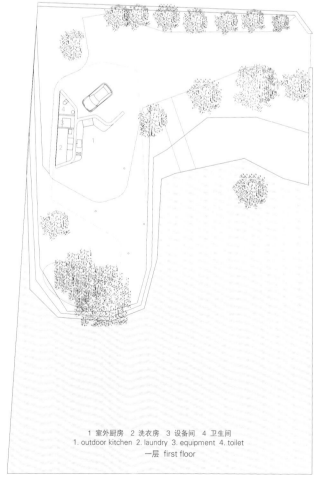

1 室外厨房　2 洗衣房　3 设备间　4 卫生间
1. outdoor kitchen 2. laundry 3. equipment 4. toilet
一层　first floor

A concrete patch traverses the plot from the lowest point up to the roof, linking the living and the inanimate. It is a patch on the landscape rather than on a board, a three-dimensional patch which absorbs the spectator into each scene, configuring interior spaces like infinitely varied enfolding curves which modify their presence according to the movement of the person inside the abode.

The new building was designed with the determination to maintain each and every element which constitutes the garden. Thus, a structure of enveloping curves appears around the swimming pool in tandem with the trees and shrubbery of the plot.

The logic of a house and a garden collapses and gives way to a totality which manages to fuse the habitable structures with the natural elements, in spite of the building's striking appearance. Every square metre of the plot is articulated so that an outer space always appears to complement each inner one. The idea of a house with its respective land has disappeared, allowing the resulting whole to be projected towards the distant landscape, the coastline and the sea with a greater intensity. An enveloping outer wall, closed off on the north from neighbouring buildings but open to the plot and the scenery, contains a second wall, which marks a subtle itinerary around the interior of the house and links each different space. People are unknowingly guided from one space to another across a fluid and continuous circular trajectory which successively introduces the different areas. The furniture of the home is adapted, arranged and attached to and around this wall, which aims to become the element around which the objects and memories of the inhabitants unfold.

Floor, walls and ceilings are concrete with different finishes, configurations, treatments and typologies: curved, flat, horizontal, vertical, perforated, smooth, polished, deactivated, trowelled, making up a catalogue of techniques and practices which employ concrete in the development of a specific project. The rich variations of the material used in its construction make this a house which is simple and complex at the same time. The building follows the concept. Only certain panes of glass momentarily turn it into a "capsule", and that is when it takes on a cosmic character.

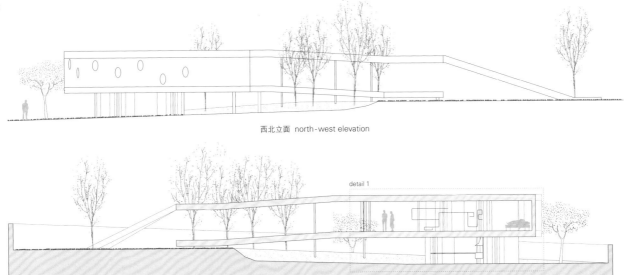

西北立面 north-west elevation

A-A' 剖面图 section A-A'

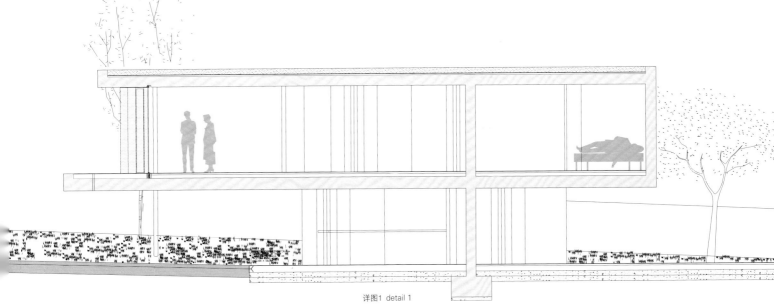

详图1 detail 1

项目名称：Family House
地点：C/Cocó, San Vicente, Alicante, Spain
建筑师：Alfredo Payá Benedito
合作建筑师：architect _ Arturo Calero Hombre, Sonia Miralles Mud, Vicente Pascual Fuentes
interior _ Beatriz Vera Payá, Natalia Velasco Velázquez, Gerardo Bernal Castell
总承包商：Construcciones Borondo
甲方：Maestre Ruiz
用地面积：1,604m²
总建筑面积：334.68m²
竣工时间：2008—2009
摄影师：©David Frutos(courtesy of the architect)

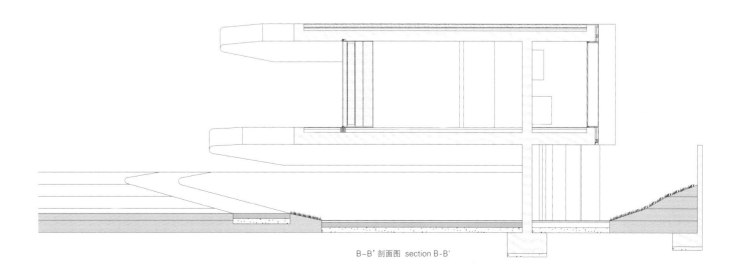

B-B' 剖面图 section B-B'

河畔俱乐部
TAO

该俱乐部坐落在盐城的一条河的一侧,周边围绕着公园和运动场。延伸的地平线、天空、水、江心岛屿以及芦苇,场地内的这些元素定义了一处宁静、纯粹且诗意的氛围。在这样的环境下,建筑师认为建筑必须是场地内的一个设计谨慎的嵌入结构,以避免破坏场地原来的感觉,同时创造与大自然的亲密接触。因此,一座位于河畔且大树环绕的玻璃建筑映入建筑师的脑海,成为最初的设计理念,以将访客、建筑和景色融为一体。

因此,设计采用密斯的Farnsworth建筑为原型理念,并通过了一系列的相关操作(拉伸、打环和折叠)创造出一种新的形式。这些操作产生了以下结果:进深小的建筑具有更好的视角,向内的庭院空间提供了更好的隐私性,更接近水和延展景观的敞开式屋顶。透明度使建筑变得非物质化。对建筑外在形态的关注被一种欲望所取代,这种欲望即创造出流动且透明的空间,使游客尽可能多地体验室外的自然环境。

为了与场地周围景观和树木的水平特征相呼应,该建筑主要为线性和折叠形式。它曲折且流动,有时接近地面,有时在空中漂浮。在建筑物里面,它在不同层面和角度都为游客提供不同的视角,同时它也给人们一个印象,即整座建筑在以很"轻"的形式接触地面,这样就产生了微妙感。因为建筑物由桩基所支撑,因此场地松软的土地条件也使得这种漂浮形式的结构变得合理。

为了强调建筑物的"轻"质特点和漂浮感,地板的厚度和柱子的大小尽量最小化。诸如低铁玻璃、白铝面板、石灰华地板、混凝土预制板和透明的玻璃隔断等材料被用来创造一种形式上的抽象感,形成简单且纯粹的气氛。在缺乏当地材料和工艺的情况下,从美学方面来说,一个抽象的形式成为了自然的选择。

Riverside Clubhouse

The clubhouse is located on one side of a river in Yancheng, surrounded by a park and sports field. The extended horizon, sky, water, island in river, and reed, these elements of the site define a tranquil, pure and poetic atmosphere. In such an environment, the architects think architecture must be a careful intervention to the site, to avoid ruining the original sense of place and meanwhile create the close contact with nature. Thus a glass building on riverside and in trees naturally comes to mind as the beginning idea, to integrate visitor, architecture and landscape.

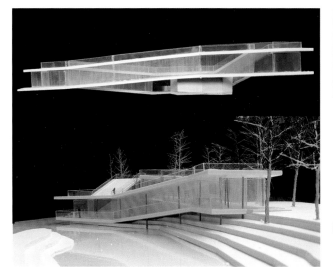

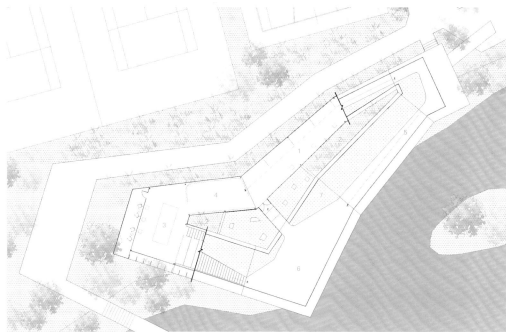

1 走廊 2 庭院 3 展厅 4 视听室 5 走道 6 屋顶露台 7 绿色屋顶
1. gallery 2. courtyard 3. exhibition 4. video area 5. walkway 6. roof terrace 7. green roof
二层 second floor

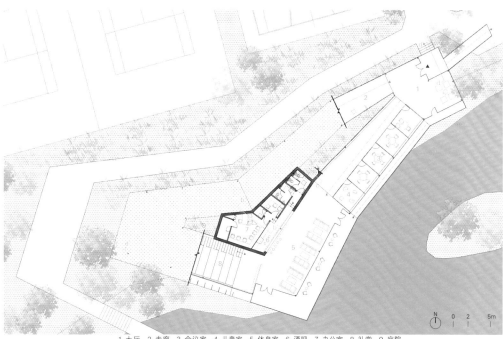

1 大厅 2 走廊 3 会议室 4 儿童室 5 休息室 6 酒吧 7 办公室 8 礼堂 9 庭院
1. lobby 2. gallery 3. meeting room 4. children's room 5. lounge 6. bar 7. office 8. auditorium 9. courtyard
一层 first floor

Farnsworth建筑为原型 Farnsworth as prototype

拉伸 stretch

打环 loop

折叠 fold

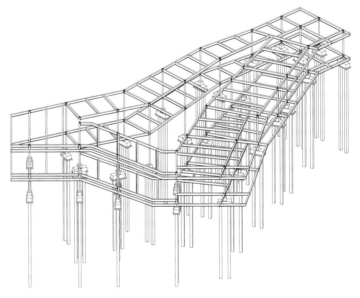

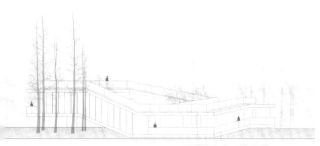
东南立面 south-east elevation

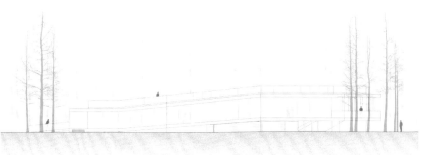
西北立面 north-west elevation

0 2 5m

1. 150mm topsoil layer
   filter fabric
   20mm plastic drainage board
   30mm fine aggregate concrete covering layer
   root-resistant waterproof layer
   75mm polystyrene-foam thermal insulation
   vapor barrier
   reinforced concrete slab
2. 3mm aluminium panel
3. 12mm low iron glass
4. white paint
   12mm gypsum board
   50/20mm u-shaped steel frame
5. 8mm+1.52mm+8mm low iron glass curtain wall
6. 140/30mm steel column
7. 30mm white travertine
   20mm screed
   reinforced concrete slab
8. 15mm precast concrete panel
   300mm reinforced concrete wall
9. 30mm light grey stone paving
   3mm hemp-fibred board insulation
   20mm fine aggregate concrete covering layer
   150mm ceramsite block
   bituminous waterproof layer
   50mm polystyrene-foam thermal insulation
   vapor barrier
   reinforced concrete slab
10. 3mm aluminium panel
    EPDM sealing layer
    40mm polystyrene-foam thermal insulation
11. 30mm steel beam
12. 10mm white ceramic tile
    20mm screed
    1.6mm polyurethane waterproof layer
    370mm fine aggregate concrete
    reinforced concrete floor slab
13. 8mm low iron glass
    12mm cavity
    6mm+1.52mm+6mm low iron glass
14. white paint
    12mm waterproof gypsum board
    EPDM sealing layer
    30mm polystyrene-foam thermal insulation
15. 30mm white travertine
    20mm screed
    reinforced concrete slab
16. 300mm steel beam

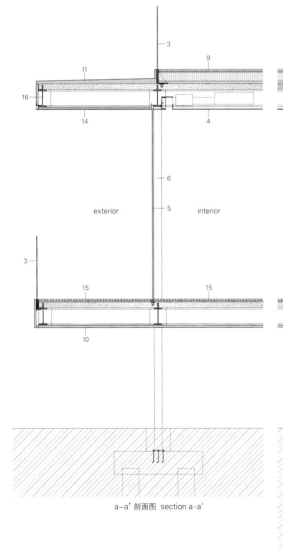

a-a' 剖面图 section a-a'

b-b' 剖面图 section b-b'

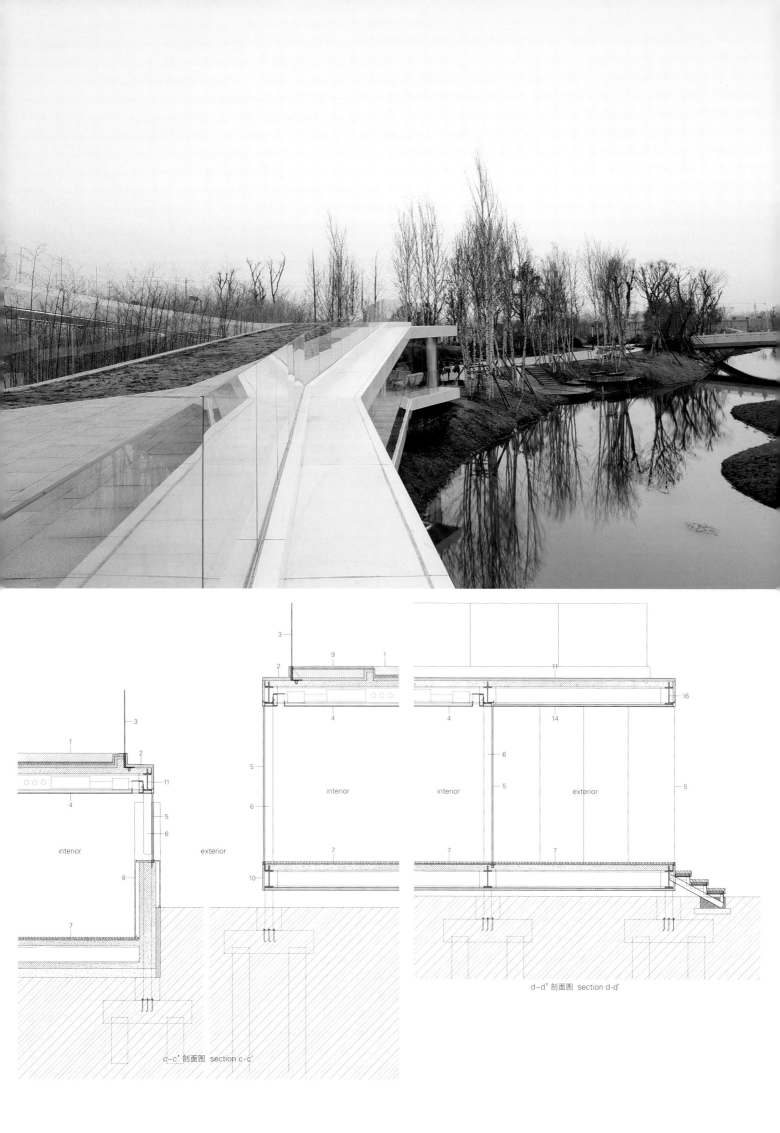

c-c' 剖面图 section c-c'

d-d' 剖面图 section d-d'

The design therefore takes Mies' Farnsworth as a prototype concept and creates a new form through a series of actions on it: stretch, loop, and fold. These actions lead to the following results: smaller building depth with better views, introversive courtyard space offering more privacy, getting closer to water and accessible roof as extension of landscape. The transparency dematerializes architecture. The concern to physical form of building is replaced by the desire to create flowing and see-through space to maximize visitors' experience of natural environment outside.
Responding to the horizontal feature of surrounding landscape and trees in site, the building is made into a linear and folded form. It zigzags and flows, sometimes approaching the ground, sometimes floating in the air. While inside it provides to visitors various views at different levels and angles, and it also gives an impression that architecture is touching the site in a very "light" form, thus creating a subtlety. The soft soil geo-condition of the site also makes this floating form structurally reasonable since slim columns on pile foundations support the building.
The floor thickness and column size are made to their minimum dimensions to emphasize the "light" character of building and feeling of floating. The materials such as low iron glass, white aluminum panel, travertine floor, precast concrete panel and translucent glass partition are used to gain formal abstraction and to create the atmosphere of simplicity and purity. In the context of lacking local materials and craftsmanship, an abstract form becomes a natural choice aesthetically.

项目名称：Riverside Clubhouse
地点：Yancheng, Jiangsu, China
建筑师：TAO
设计团队：Hua Li, Zhang Feng
结构工程师：Ma Zhigang
MEP工程师：Lv jianjun, Lian Kanglong
结构体系：steel structure on pile foundation
功能：exhibition, reception, lounge, multimedia, meeting, office
楼层面积：500m²
室外墙体材料：low iron glass frameless glazing, aluminum, precast concrete panel
设计时间：2009.11~2010.4 施工时间：2010.5~2010.10
摄影师：©Yao Li(courtesy of the architect)

Awak

苏醒的儿童空间

对于成年人来说，对建筑空间的认知主要是一种短暂的精神状态，而对于孩子来说，这种认知会对孩子意识的构建起到根本作用。正是由于这个原因，为孩子设计特别的空间就融入了真正的教育机制中来。因此，对于这方面的建筑美感与实用性的投资就升华成为对人类未来才智的投资。

超越课堂的教育空间——这种意识正在为越来越多的人所持有，自由并富于创造性的活动对于个人的教育所起到的作用与传统的学习相比正变得有过之而无不及。近期在神经科学方面的研究不仅在实验当中佐证了这种观点，而且为客观地阐释能够更好地促进孩子健康和智力发育的空间学提供了工具。

甚至在同一领域，大家仍然关注着这样一个问题——如何让孩子在不同的环境中更好地成长，以便于能够更好地适应他们的教育，从而克服由于现代的城市环境使越来越多的孩子们形成的懒惰感。因此，我们需要进行大胆且富有创造性的想象，设计出一种新的建筑类型，为孩子提供最先进的、并应用了统一标准的文化激励机制的建筑形式。

Where for adults, perceptions of a built space can be responsible mainly for certain momentary mental states, in the case of children such perceptions can contribute in a fundamental way to the formation of their mental structures. For that reason, spaces specially designed for children are true educational mechanisms, and to invest in their architectural beauty and functionality is to invest in the future intelligence of humankind.

Awareness is now growing that educational spaces go beyond classrooms, and liberal creative and physical activities contribute to the individual's education at least as much as traditional learning does. Recent studies in neuroscience not only empirically confirm this awareness, but provide tools for objective definitions of spaces better able to maximize children's wellbeing and intellectual growth.

Even at that, a concern remains as to how well young people raised in environments thus well adapted to their education will be able to cope with the increasing sloppiness of contemporary urban environments. A great creative effort is therefore called for to imagine a new range of building types that observe the same criteria for cultural stimuli applied in the most advanced forms of architecture for children.

Learning from Spaces

Young people's cognitive responses to a built space have far greater consequences than the similar perceptions of normal adults given that the brains of children between the ages of 6 and 12 are still undergoing major anatomical changes.[1]

Where the spatial perceptions of an adult's fully developed mind occasion momentary discomfort or an enhanced sense of wellbeing, in the still evolving mind of children those perceptions contribute to a great extent to determining their qualities and abilities throughout life.

Influenced by the changing organization of contemporary society and the abandonment of traditional family structures, the time young people spend in the family environment has been reduced to a few moments each day, with this time often dedicated to television, the Internet or video games. In such reductive conditions, the quality of domestic space and time, unfortunately, plays an increasingly marginal role in young people's development.

Ideally, the developmental needs of children should be addressed not merely through systems of teaching, but through introduction to complex environments which offer intellectual, sensory and motor stimulation, thus facilitating the construction of strong self-confidence and the development of cognitive skills.

The physical spaces with which children interact, whether school classrooms, playgrounds or playrooms, are much more than empty containers: They constitute an essential source of information, like text books or audiovisual media.

Such findings, long supported by logic and intuition, take on a deeper meaning today given how they have been strengthened and made more operational by scientific evidence.

As John Paul Eberhard, founding President of the Academy of Neuroscience for Architecture, wrote,

"Research undertaken by neuroscientists around the world is beginning to provide new insights into the influence of the various qualities of schools on learning experiences. Schools designed with an understanding of how children's minds respond to the attributes of spaces and places can lead to enhanced learning. Such research is adding to the architectural knowledge based on an understanding of daylight,

奥伦赛幼儿园_Kindergarten in Orense/Abalo Alonso Arquitectos
Saunalahti幼儿园_House of Children in Saunalahti/JKMM Architects
国王公园环境意识中心_Kings Park Environment Awareness Center/Donaldson+Warn
Lucie Aubrac学校_Lucie Aubrac School/Laurens&Loustau Architectes
Ama'r儿童文化馆_Ama'r Children's Culture House/Dorte Mandrup
Lasalle Franciscanas学校高架运动场_Lasalle Franciscanas School Elevated Sports Court/Guzmán de Yarza Blache
从空间中获取知识_Learning from Spaces/Aldo Vanini

从空间中获取知识

年轻人对建筑空间的认知反应要比普通成年人的相似类型的认知反应产生更为深远的影响，而我们还要考虑到六至十二岁的孩子的大脑结构仍然处于重大的变化之中。[1]

对于成年人已经发育成熟的思维来说，空间上的认知意味着偶尔短暂的不舒服或者是提升了的舒适的感觉，但是对于仍然处于思维进化中的孩子来说，这种认知会在很大程度上决定他们日后所具备的素质和能力。

受到当代社会不断变化的组织结构和对传统家庭模式的抛弃的影响，孩子每天待在家里的时间已经减少到一定的程度，而这仅有的一点时间也经常被用来看电视、上网或者是玩电子游戏。不幸的是，在这种家庭时间日益减少的情况下，家庭里空间和时间的质量对于年轻人的成长也正起着越来越微不足道的作用。

从理想化的角度来说，孩子的成长不仅需要相应的教育体系，还要通过引入复杂的环境来实现，因为环境可以向孩子提供智力上、感知上和行动上的激励，从而促进孩子强大的自信心的建立和认知技能的发展。

孩子所处的实际互动空间，不管是教室、操场还是游戏室都不应当是毫无一物的空间：这些地方应当如教科书或者视听媒体等方式一样成为信息的基本来源。这些长期以来由逻辑和直觉所支持的发现，在现代科学证据的进一步佐证下变得更加容易实施，在今天，这些发现也呈现出了一种更深的含义。

正如约翰·保罗·艾伯哈德——建筑科学学院创始人兼总裁写的那样，"在整个世界范围所展开的精神科学研究为学校的不同特质对学习经验所产生的影响提供了新的视角。学校的设计应该着眼于理解孩子的思维对于空间特征的反馈以及空间对于学习的提高作用上。这样的研究是基于对光线、声学、空气质量以及对于自然的观点这些能够深深地影响孩子的认知进程的因素的理解上，并赋予到建筑知识的。"[2]

目前，为孩子们设计的设施已经从相对简单的设计发展为对更加大胆的概念的尝试，过去的设计主要起源于19世纪，主要以对包括幼儿园、学校和操场的环境的设计为特点的。

当代的设计主要把重点放在对内部与外部的交流通道的设计上，并建立在对空间布局的反馈和激励导视能力的研究上，致力于提高思维与外部环境的关系这种基本的能力。这种多感官的激励不仅可以对于具有正常技能的孩子，而且对于有一定残疾的孩子，在将空间构建和融合为现实的领悟方面有很大的帮助作用。传统的正交性主要是出于经济实用的考虑而产生，并没有更多复杂的指向与组合，这迫使孩子的思维进入一个他们正在践行却需要不断更新的空间模式。给年轻人的生活注入一些情景化的地标非常重要，包括颜色、图像或有趣的物体，可以帮助构建空间记忆。此外，基于最近的神经科学实证的研究表明暴露在自然光下学习效果更好。儿童设施的现代化设计趋向于强调光源的作用，而这不仅仅是简单的方形窗户，往往是任意导向的。

大量关于建筑环境和大脑各部分之间关系的科学研究仍有待完成，诸如光、视觉模式、记忆与认知过程等方面。然而，将艺术性与创造性的设计应用在简单而机械化的工程中所产生的巨大价值不容小觑。

以下例子体现了一些为大多数孩子成长过程的教育环境实行最新标准的考虑。

*acoustics, air quality, and views of nature that deeply affect the cognitive processes of children."*²

In recent times facilities for children have evolved from the relatively simple designs that formerly characterized such environments beginning in the XIX century, including those for pre-schools, schools and playgrounds, into more daring concepts.

In such contemporary designs, special emphasis is placed on internal and external paths, based on studies of responses to spatial arrangements and of the stimulation of wayfinding ability, a fundamental capacity in enhancing the relationship of the mind with its surrounding world. Such multisensorial stimulation assists normally skilled children as well as those suffering from certain disabilities in constructing and integrating spatial representations into their global interpretation of reality. Traditional orthogonality, which emerges primarily from economic and practical considerations, gives way to more complex directionality and composition, forcing the child's mind into a continual reworking of the spatial model in which he or she is acting. It is also very important to give the youngsters living in these contexts landmarks, including colors, images or interesting objects, to help in the construction of spatial memory. Furthermore, based on recent neuroscientific empirical demonstrations of how learning may be enhanced by exposure to natural daylight, contemporary designs for children's facilities tend to emphasize the role of light sources that are more than simple squared windows, often casually oriented.

Much scientific studies remain to be done into the relationship between the built environment and the developing and learning brain, including such aspects as light, visual patterns and memory and cognitive processes. However, the tremendous value that artistic and creative architectural design adds to simply and mechanically functional projects must never be forgotten.

The following examples embody all these considerations in putting into practice the most up-to-date criteria for educational environments that oversee the vast majority of a child's developmental process.

Conceived as a facility for the University Campus of Orense, Spain,

Saunalahti幼儿园。其操场在一处难以开发的崎岖场地内形成一处安全无阻且充满欢乐氛围的人工景观
House of Children in Saunalahti. The playground forms a safe, unobstructed and exhilarating artificial landscape at a difficult and rocky site.

Abalo Alonso建筑事务所接手了在西班牙奥伦赛大学校园里建一个托儿所的项目,其灵感来自于城镇周边林区的小木屋和传统式粮仓。为了顺应当地倾斜的地势,此建筑被设计成坐落在混凝土基座上的木质房屋,正是得益于这种来自然又回归自然的木质材料,它召回了木屋的自然姿态,让孩子们欢快地嬉戏于此。简易的矩形周界由一些薄薄的木质材料围成,形成一个可调节的自然日光过滤器,而庭院里的日光补充了这里光线的不足。植物打破了非直角相交空间的传统模式,无论内部还是外部都能持续刺激孩子的感知力,这种外观的随意性有助于孩子们重建空间模式。整个托儿所的空间一直延伸到一个开放式的、无栅栏的草地。

建筑师Guzmán de Yarza Blache为西班牙萨拉戈萨的Lasalle Franciscanas学校设计的高架运动场是一个露天的功能性建筑的典范。这不仅是孩子们的游乐场,更是家长接送孩子的集散地。为了避免多项活动同时进行的不便,建筑师考虑升高水泥柱之上的金属结构。这样一来便产生了三种重要的结果:隔开了各种游戏与社交活动,建成一个色彩鲜艳的覆顶广场,并且在不破坏学校庭院原有风貌的基础上形成一个覆盖多种常青藤的地标。这种新颖的结构迅速被学生们所接受并被亲切地称为"鲸鱼"。

Laurens&Loustau建筑师事务所为西班牙图卢兹的Lucie Aubrac学校设计了传统的正交组合的矩阵布局,其创新之处在于将外部空间与室内教学活动融合成一体。厚厚的混凝土使这个室外活动场变得安全,人们通过圆形洞口可以看见天空。此设计大量使用色彩来结合水平表面和垂直表面,天花板被分解成多层次面板的序列形式。同奥伦赛托儿所的建筑材料相似,该教育综合体垂直的薄木条像厚窗帘一样过滤了来自外面的光,因此这里十分注重光照。

多特·曼德鲁普在她的Ama'r儿童文化馆中测试了与结构相关的教育理念,她还测试了她的设计和历史的市镇之间的关系的局限性。她的文化屋被想象和设计成大山一样的建筑,它从它所处的环境中吸收了城市群的形态学理念,并打上了深深的烙印,文化馆背离普通的建筑因素的等级评定,使自身构造的模糊性与周围环境构造的极端严格的标准形成反差。这其中的一个例子就是建筑的外部是由玻璃覆盖的,看起来在横向和纵向上毫无差别,而且这种设计看起来更像是向行人展示内部奇幻般的世界,而不是只为里面的人提供一个向外观察的通道。内部的设计也与这种逻辑相呼应,它让人们打破已形成的观念,转而向年轻的

the nursery planned by Abalo Alonso Arquitectos was inspired by the wooden cabins and horreos, the traditional granaries, found in the woodland surrounding the town. To follow the sloping terrain, the building is designed as a wooden object resting on a concrete pedestal, so that it recalls, thanks to a choice of materials either natural or evocative of a natural attitude, the small tree houses so dear to childhood play. Its perimeter, elementary and rectangular, is defined by a sequence of thin wooden elements which constitute a modifiable filter of the natural daylight, supplemented by that from internal courtyards. The plant breaks the traditional pattern in being crossed by non-orthogonal spaces, alternately internal or external, that offer the child continuous perceptive stimulation. Its apparent randomness is governed by the presence of color patterns that help children reconstruct the spatial model. The nursery extends to an open, unfenced meadow.

lasalle Franciscanas School Elevated Sports Court in Zaragoza, by Guzmán de Yarza Blache, is an example of open-air functional intersection. More than just a playground, it is a public plaza, a collection and departure point for parents. To avoid the inconvenience of overlapping activities, the architects have elevated the metallic structure on concrete pillars with three important results: to separate the various play and social activities, to create a covered plaza enlivened by vivid colors, and to establish a sort of landmark that does not spoil with its imposing presence of the ancient school courtyard, covered by many kinds of ivy. This new structure the students promptly nicknamed "the whale".

Laurens & Loustau Architectes set out a more traditional, orthogonal compositional matrix for the layout of the Lucie Aubrac School, focusing their innovative intention on bringing the external space into complete continuity with indoor creative and educational activities. A thick concrete "cloud" protects the outdoor playground, with the sky visible through circular openings. A generous use of colors integrates horizontal and vertical surfaces, while the ceilings are broken down into sequences of multi-level panels. Special attention is paid to lighting, with the indirect light coming from above sometimes filtered by a thick curtain of thin vertical wooden splints, similar to those used in the Orense Nursery.

Dorte Mandrup, in her Ama'r Children's Culture House, tests the limits both of educational concepts related to structures and of the relationship between her design and the historic town. A container conceived and organized as a mountain, the Culture House accepts from its context only the morphological footprint of the urban block as it develops into a volume that breaks with the usual hierarchy of architectural elements, setting its tectonic ambiguity against the extreme compositional rigor of surrounding structures. One example among several of this opposition is the windows not only cover the outer surfaces without observing a distinction between verticality and horizontality, but seem designed more to show passers-by the magical world within than to serve as an observation portal for those inside. The interiors

照片提供：©Laurens&Loustau(Stéphane Chalmeau)

Lucie Aubrac学校的穿孔混凝土屋顶使这个室外活动场变得安全，人们通过圆形洞口可以看见天空。
The punched concrete roof of Lucie Aubrac School protects the outdoor playground, with the sky visible through circular openings.

1. Halfon N., Shulmam E., Hochenstein M., "Brain development in early childhood", *Building Community Systems for Young Children*, Los Angeles: UCLA, 2001.
2. Eberhard J.P., *Brain Landscape*, New York: Oxford University Press, 2009.

使用者们展示了一系列神奇的因素——形状、颜色、意象和方向——所有这些使他们必须不断地去解构这个不同寻常的空间设置，通过这样又可以使他们对于现实构造的理解取得不凡的进步。从婴儿期到十八岁，对于孩子的进化中的思维来说，将一切事物想象成一处奇幻的空间对于他们的想象力是尤其关键的。

坐落于澳大利亚珀斯市最富有的公园中，由道森+沃恩建立的国王公园环境意识中心不像是一个美学结构，而更像是一个致力于可持续发展的项目。通透的结构与周围自然的环境相得益彰，土质的屋顶从上面遮掩着各种设施。一条蜿蜒的走廊其实就是公园里的一条步行道，它直接通向内部的小广场，而那里就是他们的工作室，里面不同功能的空间也只是通过大的玻璃屏分开。事实上，这个项目的首要任务是要最大程度地实施环境促进可持续发展的各种措施。建筑物的教育方面的内容则主要是以使用者和公园的自然因素的接触为前提的，并通过使用最少的建筑因素实现的。

受到来自于传统的诺蒂克极简抽象派的启迪，由JKMM建筑师事务所设计的芬兰埃斯珀Saunalahti幼儿园是一个日间护理中心，提供两个等级的不同功能的服务。它在儿童教育空间的主题方面使用了独具一格的方法：通过使用与欧洲的远北地区文化相符合的理念，它的建筑表现出一种非常宁静的环境，与之相反的是通过之前的实例来产生爆炸性的感官上的激励作用。朝向来路的这座曲形建筑的立面有一个别名——Lowly Worm，是以理查德·斯佳丽创作的科幻人物的名字命名的，它通过看起来凌乱的开放式体系打破了传统的理念，完全展现在坐落于包围着村庄的森林旁的综合性大操场面前。建筑的房顶突出了现代运动类型的、怀旧的大天窗的特点，并在一些窗户的隔栏里装饰着动物的图案。

现在就存在着一个亟待解决的问题。今天，教育已经被普遍认为是一个持续一生的过程，正是由于这个原因，从孩童时期开始的所有学习阶段都值得我们投以最大的关注，同样地，我们还有关注于空间对于教育的潜能的发挥所产生的有意识或无意识的、明确的或者是隐含的作用。在高素质教育的环境中成长起来的孩子日后一定会继续在同样标准条件下设计和建造的设施当中，拥有个人与社会发展的机会。考虑到新形式的社会组织将需要付出大量的努力来设想适应新功能的空间类型，这一要求也一定会被看作是整个新领域建筑类型发展的一个的关键契机。而这一全新并富于创造性的挑战将首先从儿童设施的设计革新开始又是多么的恰如其分啊。

echo this logic in breaking from preconceived schemes to expose young users to a series of surprising elements – shapes, colors, images, directions – that force them into a continuous encoding of unusual spaces in a way that promises extraordinary growth in their understanding of the structures of reality. Everything is conceived as a magical space that is particularly congenial to the imagination of the evolving minds of young people from infancy to 18.

Situated in the most treasured park in Perth, Australia, the Kings Park Environment Awareness by Donaldson+Warn is oriented less to an aesthetical than to a sustainable program. The transparent shelter is in continuity with the surrounding nature, located under an earth roof that conceals the facility from above. A zigzagged corridor that is part of the park trail network leads to an indoor plaza, an atelier, and various functional spaces separated from the context only by large glass pans. Indeed, the project's priority is to implement environmental sustainability measures of the highest level. The educational content of the building, in fact, is premised upon maximum contact between users and the park's natural elements, via minimal use of architectural elements.

Inspired by traditional Nordic Minimalism, the House of Children in Saunalahti, Espoo, Finland, by JKMM Architects, accommodates on its two levels the various functions required of a day-care center. It represents a different approach to the theme of children's educational spaces: With an attitude suited to the cultures of Europe's Far North, the building presents itself as an environment of great serenity, in contrast to the explosive sensory stimulation of the previous examples. The road-facing facade of the sinuous volume, nicknamed Lowly Worm after Richard Scarry's fictional character, is pierced by a seemingly random system of openings, and opens fully onto a large and complex playground that borders the forest surrounding the village. The roof is marked by large skylights reminiscent of Modern Movement examples, in some cases decorated inside with animal images.

One pressing question remains to be addressed. Today, education is understood as a process that is ongoing for a lifetime. For this reason, all the phases of learning from childhood on deserve our deepest attention, no less so than the educational potential of the spaces within which this process takes place, knowingly or unknowingly, explicitly or implicitly. Children raised in spaces of high educational quality must have the opportunity to continue their personal and social growth in facilities designed and constructed according to similar criteria. This requirement must be thought of as a key opportunity in the development of an entirely new range of building types, in view of the fact that new forms of social organization require great effort in imagining new spaces appropriate to new functions. It is fascinatingly apt that this new creative challenge should begin with innovations in the design of children's facilities. *Aldo Vanini*

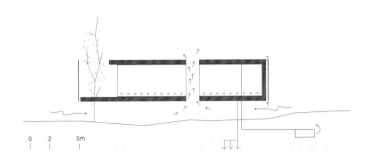

奥伦赛幼儿园

Abalo Alonso Arquitectos

该项目可能源于对加利西亚粮仓的记忆，从一些林间木屋的形象得来，但场地的实际布局最终占据了中心舞台。

出于使用性和可行性的原因，该幼儿园占据一个单独楼层。入口处位于场地的最高点，引导人们利用吊桥来进入一栋建筑。这座建筑根据功能划分为五个平行的带状区。多功能大厅对空间结构起到了规划作用：北面是带有独立入口的服务区、办公室、员工更衣室、厨房和设备间。南面是带有衣柜和盥洗室的滤光装置、教室和带有游乐区的阳台，占据了奥伦赛大学校园。教室之间的柔软的围护结构允许产生不同的布局，从完全地独立于彼此到开放式房间。尽管舒适的程度不同，但非连续的楼层平面、一些玻璃窗格和镜子的反射以及内部和外部相似的建筑材料的利用都有助于增加其丰富性。

该建筑是由四个粗荔枝面的混凝土构件所支撑。这些混凝土构件嵌入了供水、雨水收集、卫生、电力、电信和地热能方面的设备。两个水平向混凝土板构成地板和天花板，使结构钢筋最为必要，使人们想起加利西亚粮仓的支柱。建筑物的相对位置使侧面形成具有足够高度的开放式室内空间，这处空间既可在雨天使用，也可在平时炎热的夏天使用。

建筑的立面围护结构由19cm厚的黏土块制成，内侧的保温层则为一个包括地板和天花板的连续层，意在消除热量流失。屋顶铺满了黑色板岩片，放置在几何线条之间，重新打破内部的区域划分。

作为覆盖材料，木材的选择必须面对这样的事实，即材料的重量要十分轻，这都是该项目中用于防止悬臂结构超负荷的重要考量因素。因此，建筑的最终图像是个太阳伞的形象，并且形成一个通风过滤器，来减少照射在建筑物上的日光量及其能量消耗。在另一个方向上的教室的朝向有利于捕获冬季里的阳光。根据露台的实际条件可以考虑在夏季安装遮阳篷，以保护儿童游乐区。阳光与格栅一起配合上演了一出光与影的迷人剧目。雪松材质的格栅覆有彩色的保护层，将整个格栅包裹起来，赋予这座建筑最终的形象，并充当遮阳伞，来帮助调节不同洞口的盲区。地面上的洞口，尤其是中央庭院，有助于在立面之间，甚至在其与楼层之间产生对流通风，从而减少了夏季人工能耗。

Kindergarten in Orense

This project probably arose from the memory of the Galician hórreos (granaries) and from the image of some huts among the trees, but the actual configuration of the land ultimately takes centre stage.

This nursery school occupies a single floor for reasons of use and accessibility. The entrance, situated at the highest point of the land, leads people in the manner of a drawbridge into a building functionally organized into five parallel strips. A multifunctional hall organizes the spatial structure: towards the north the server spaces with their own independent access, office, staff changing room, kitchen and installation room. To the south is a filter with wardrobes and lavatories, classrooms and balcony with play area, dominating the university campus. The flexible enclosures between classrooms permit different layouts that go from total independence from each other to the possibility of an open-plan room. The breaks in the floor plan, together with the reflections of some glass panes and mirrors and the use of similar materials inside and outside the building, favour the feeling of amplitude despite its snug size.

The building is supported by four bush-hamered concrete screens that have embedded in them the installations for the water supply, rainwater collection, sanitation, electricity, telecommunication and geothermal energy. Two horizontal slabs, also

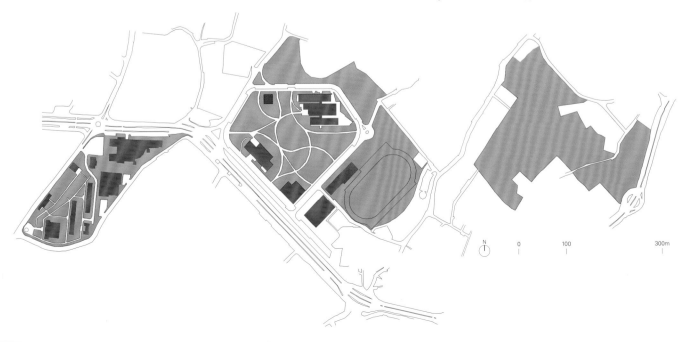

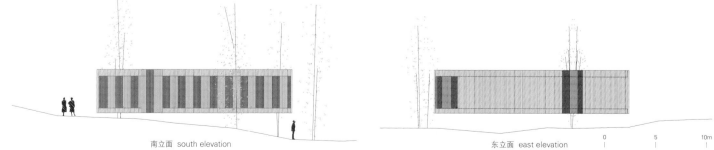

南立面 south elevation   东立面 east elevation

A-A' 剖面图 section A-A'   B-B' 剖面图 section B-B'

项目名称：Kindergarden, University Campus, Orense, Spain
地点：University Campus, Orense, Spain
建筑师：Elizabeth Abalo, Gonzalo Alonso
合作建筑师：Berta Peleteiro
结构工程师：Carlos Bóveda
设备工程师：Fernando Gago
地热公司：Xeoaquis
技术建筑师：Francisco González Varela
开发商：Vigo University, Orense University
施工单位：Construcciones Paraxe
场地管理：Fernando Soengas
总楼面面积：327m²
竣工时间：2011
摄影师：©Santos-Díez | BIS Images

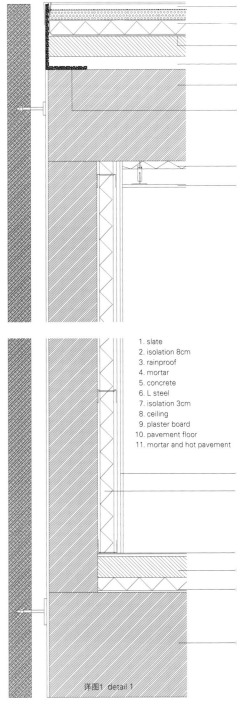

1. slate
2. isolation 8cm
3. rainproof
4. mortar
5. concrete
6. L steel
7. isolation 3cm
8. ceiling
9. plaster board
10. pavement floor
11. mortar and hot pavement

详图1 detail 1

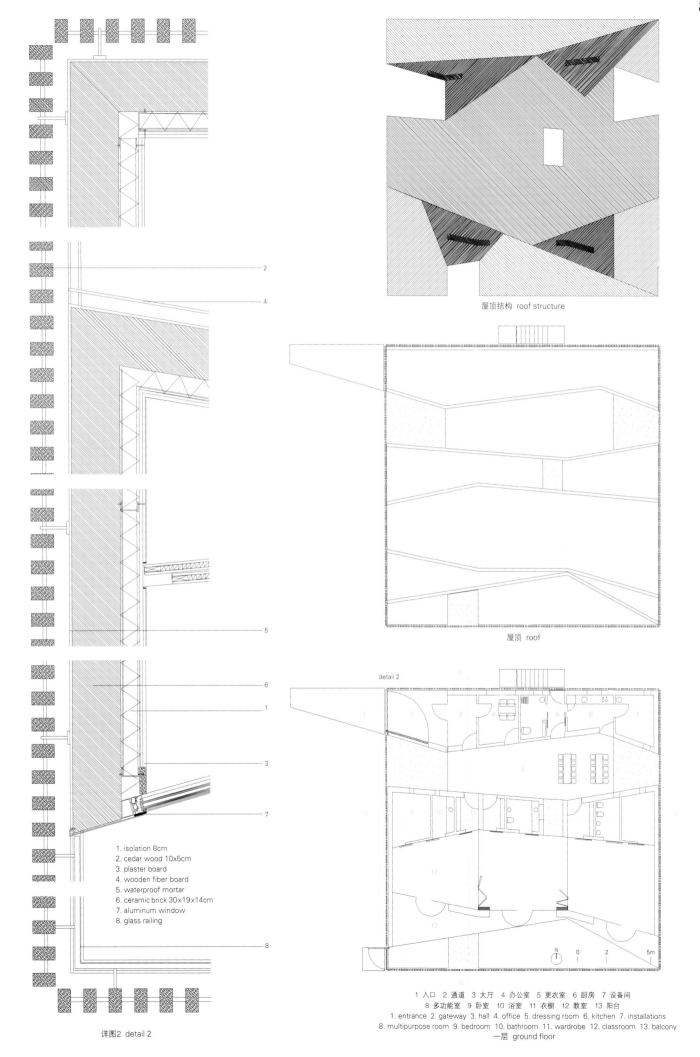

1. isolation 8cm
2. cedar wood 10x5cm
3. plaster board
4. wooden fiber board
5. waterproof mortar
6. ceramic brick 30x19x14cm
7. aluminum window
8. glass railing

详图2 detail 2

屋顶结构 roof structure

屋顶 roof

1 入口 2 通道 3 大厅 4 办公室 5 更衣室 6 厨房 7 设备间
8 多功能室 9 卧室 10 浴室 11 衣橱 12 教室 13 阳台
1. entrance 2. gateway 3. hall 4. office 5. dressing room 6. kitchen 7. installations
8. multipurpose room 9. bedroom 10. bathroom 11. wardrobe 12. classroom 13. balcony
一层 ground floor

in concrete, form the floor and ceiling, making the most of the necessary structural reinforcements that recall the supports of the Galician hórreos. The relative position of the building generates in the section an open plan of interior space with sufficient height that can be used both on rainy days and, above all, on the usually hot summer days.

The facade enclosure is made from clay blocks of 19cm thick and the insulation goes in the inner side in a continuos layer, including floor and ceilings, with the intention of eliminating any thermal breaks. The roof is crowned with black slate chips placed between geometric lines to reproduce the broken interior divisions.

The choice of wood as the cladding material has to do with the fact that it has little weight, something that is important in this project to prevent overloading the overhangs. Thus, while resolving the building's final image, it collaborates by acting as a parasol, generating a ventilated filter that reduces the amount of sunlight that strikes the building as well as its energy consumption. The orientation of the classrooms in the other direction favours the capturing of sunlight in winter. The situation of the terrace allows awnings to be installed in summer to protect the children's play area. Together with the lattice, it provides an attractive play of light and shadow.

A cedar wood lattice treated with a coat of "lasur" envelops the entire piece and gives the building its final image, helping to regulate it by acting as a parasol of the blind part of the different openings. The ground floor openings, and above all the central courtyard, facilitate the generation of crossed ventilation not only between facades but also with the floor, thus reducing artificial energy consumption in the summer.

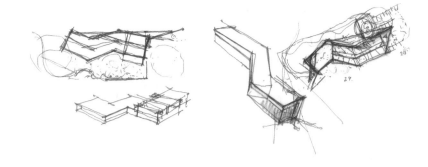

Saunalahti幼儿园
JKMM Architects

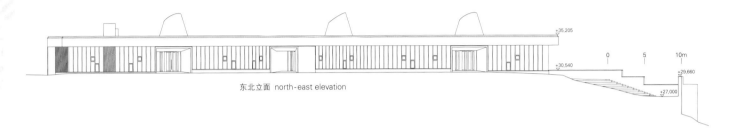

东北立面 north-east elevation

这所幼儿园的参赛作品的主题是"马托·马塔拉"(《卑微的蠕虫》，理查德·斯卡里)。二楼设有日托中心，包含五组不同年龄阶层儿童的活动空间和公共空间。日托生活区域在覆盖着松树的山坡和新建筑之间的游戏场上展开。街边包含一系列常用设施。一楼包括托儿所和技术控制室。

弯曲的南墙形成了建筑的公共立面。立面的其余部分均采用木材建造。大厦坐落于Saunalahti海鸥海滨附近崎岖的岩石地区。操场上形成了一个安全、通畅并且使人为之一振的人工景观。建造该景观的动机、材料和色彩的灵感均来源于现存实景。

建筑物的主体结构是用混凝土制成的。西南立面由接缝的轻砖来建造。其他的立面都贴满了木框窗户。天窗向下面的每个单元的入口大厅开放，可饱览大海、陆地和空间的景色。室内装饰材料采用了木灰和填料地板、绿色簇绒地毯、应用在墙壁上的灰泥和天花板上的反光隔音涂膜纸。特殊的灯饰配件和固定装置以及几件家具是为这座建筑单独设计和定制的。此外，大楼的设计旨在在孩子们玩耍的世界里掺杂进童话影像。

House of Children in Saunalahti

Motto of the competition entry for House of Children was "Mato Matala" (Lowly Worm by Richard Scarry). Second floor houses day-care centre containing five groups of children and common spaces for all users. The day-care home areas open out on the playing yard that is formed between the rising pine covered hill slope and

西南立面 south-west elevation

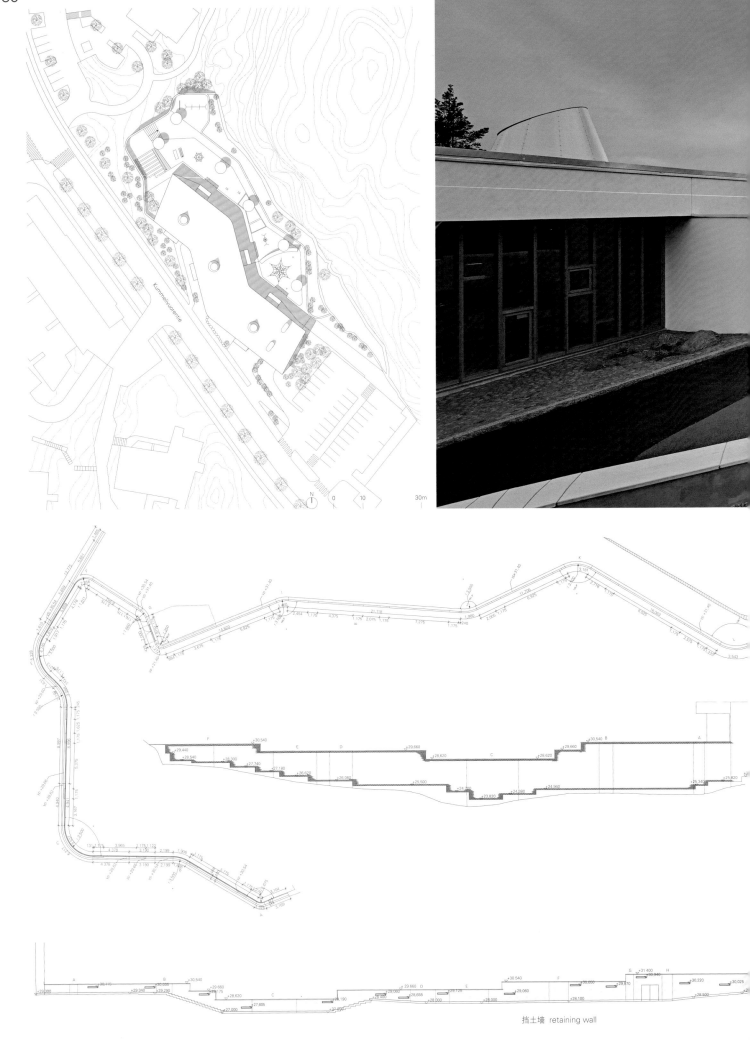

挡土墙 retaining wall

the new building. The street side contains common and staff facilities. First floor contains children's nursery and technical spaces. The curved southern wall forms the public facade of the building. Rest of the facades are made of timber. This building is located at a difficult, rocky site near Saunalahti Gulf Seashore. Playground forms a safe, unobstructed and exhilarating artificial landscape. Motives, materials and colours of the building have been inspired by the excisting landscape.

The main structure of the building is made out of concrete. Southwest facade is light masonry with overspread joint sealing. Other facades are plastered with wooden frame windows. Skylight windows open down to the entry hall of each unit and contain images from sea, earth and space. Interior materials are wooden ash parquettes and filler floorings, green tufted carpet and plastering on the walls and light acoustic papercoating on ceilings. Special light fittings and fixtures as well as pieces of furniture are individually designed and customized for this building. The design of the building is intended to stir fairytale images in the world of playing for children.

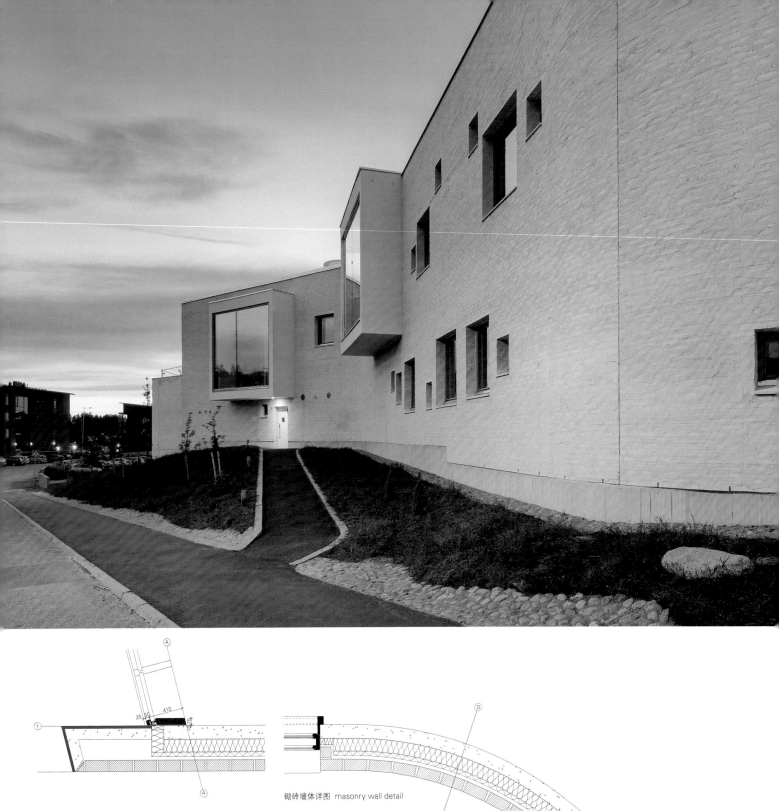

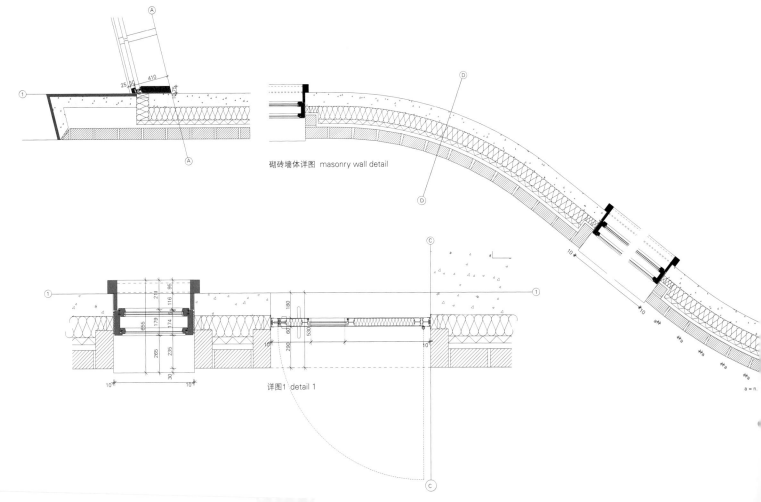

砌砖墙体详图 masonry wall detail

详图1 detail 1

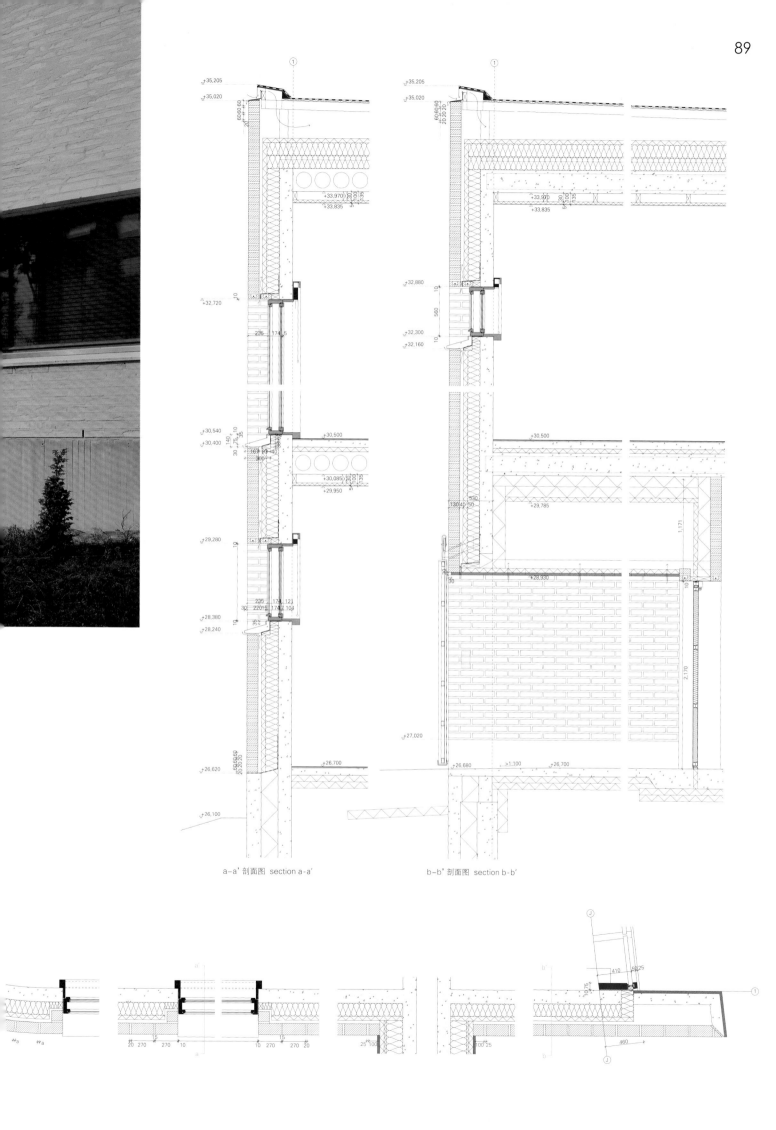

a-a' 剖面图 section a-a'   b-b' 剖面图 section b-b'

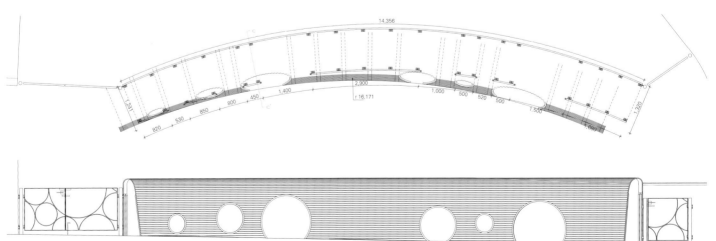

游乐隧道围墙详图 playing tunnel fence detail

项目名称：House of Children in Saunalahti, Espoo
地点：Kummelivuorentie 2, 02330 Espoo, Finland
建筑师：JKMM Architects
项目团队：author_Samuli Miettinen, Asmo Jaaksi, Teemu Kurkela, Juha Mäki-Jyllilä /project architect_Katja Savolainen/project architect during construction phase_Christopher Delaney /interior architect_Päivi Meuronen /Edit Bajsz, Aaro Martikainen, Ilona Palmunen, Merita Soini
结构工程师：Finnmap Consulting ltd
服务设计：Livair ltd
电气和照明工程师：Tuomi Yhtiöt ltd
景观建筑师：Loci Landscapearchitects
技术家：Aimo Katajamäki, Ilona Rista
总承包商：Rakennuskartio ltd
甲方：City of Espoo
用途：daycare center, nursery
用地面积：5,522m² 总建筑面积：2,013m² 体积：6,300m³
设计时间：2007.6 竣工时间：2011.7
摄影师：©Mika Huisman(courtesy of the architect)(except as noted)

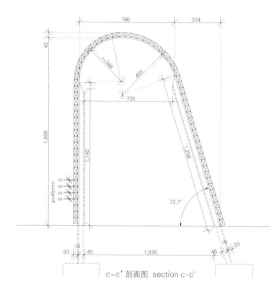

c-c' 剖面图 section c-c'

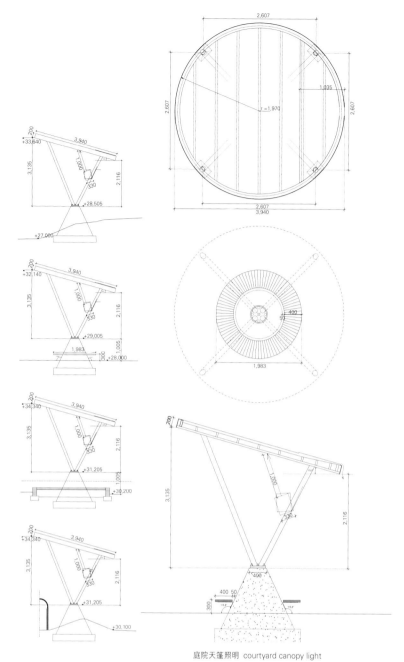

庭院天篷照明 courtyard canopy light

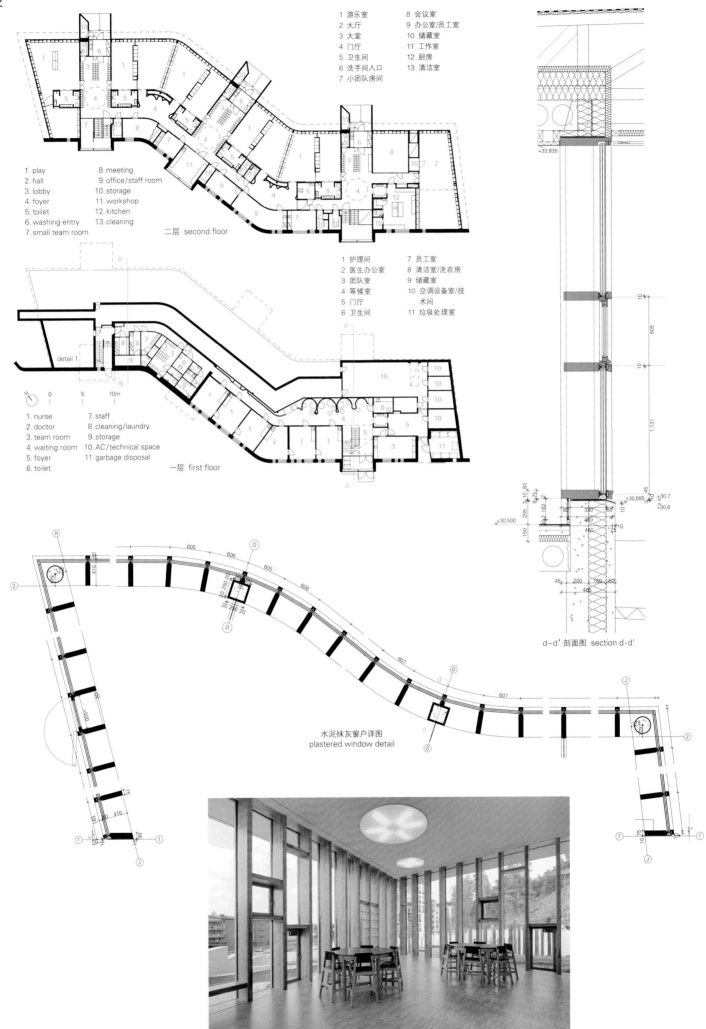

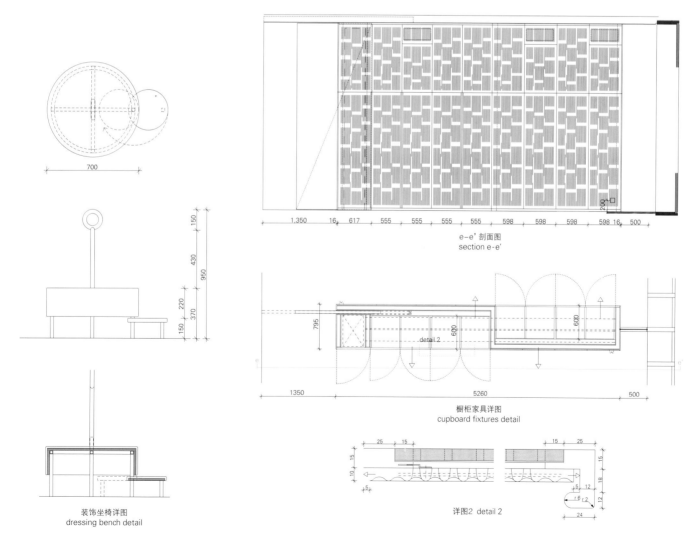

e-e' 剖面图
section e-e'

橱柜家具详图
cupboard fixtures detail

装饰坐椅详图
dressing bench detail

详图2 detail 2

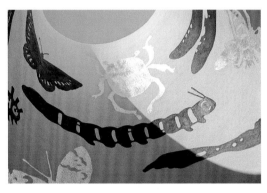

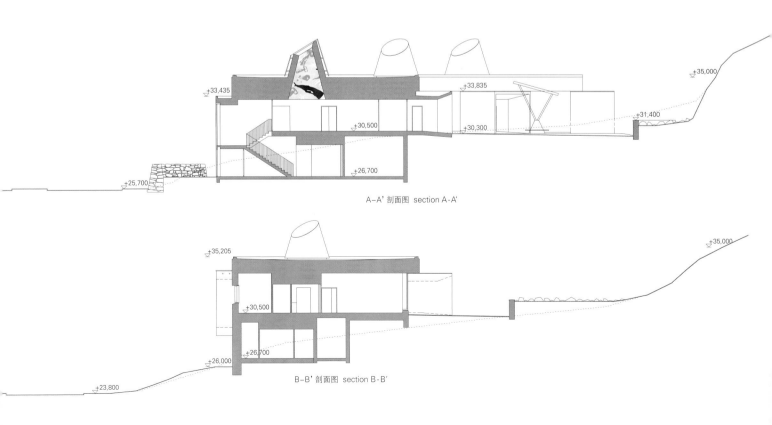

A-A' 剖面图 section A-A'

B-B' 剖面图 section B-B'

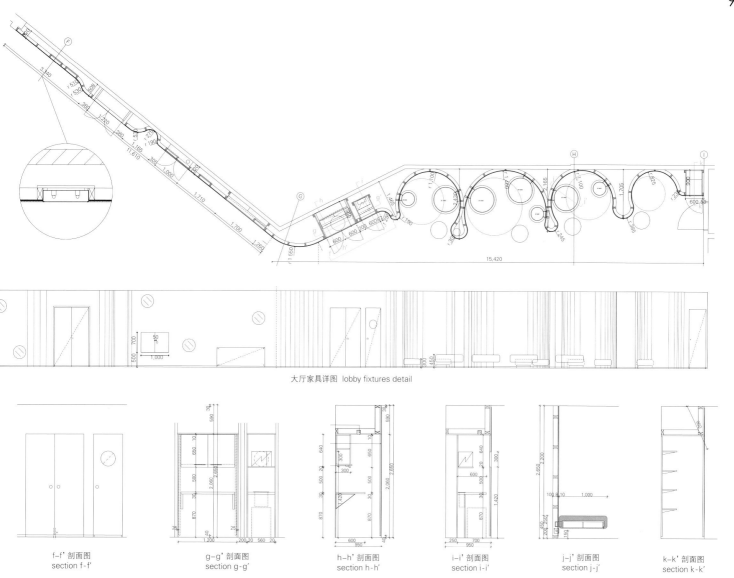

大厅家具详图 lobby fixtures detail

| f-f' 剖面图
section f-f' | g-g' 剖面图
section g-g' | h-h' 剖面图
section h-h' | i-i' 剖面图
section i-i' | j-j' 剖面图
section j-j' | k-k' 剖面图
section k-k' |

国王公园环境意识中心
Donaldson + Warn

　　国王公园是澳大利亚珀斯市最珍贵的公共花园,因而园内的一座新建筑也是一个重要的委托项目;作为自然景观管理区的一部分,国王公园环境意识中心是植物园及园林管理局实施教育方案的服务机构和大本营,虽然这座建筑主要为学龄儿童使用,但是也会被用于公共展示及其他功能。

　　这座富于表现力和体验性的建筑有一个关于可持续发展的重大议程,集中在采用太阳能发电、降低能源消耗,同时与人们分享可持续发展的美学表现。设计师将建筑本体、所在场地及人工覆土后重新长出天然植物的屋顶折叠在一起,借此设计来对应表现人类、居所及自然环境间的先后顺序关系。地方植物群的主角形象通过三种表现手法得以强调:植被覆盖的屋顶景观;渐渐沉入土中的长长的入口坡道,将游客引到一处潜入灌木丛中的露天平台;为使用者提供的一个观看外景的、全景视野的落地式双层玻璃墙。翻光面呈弧形环绕着建筑的轮廓,充分发挥了良好的朝向带来的优势,而蓄热体及双层玻璃、LED照明、再生材料、对流通风和太阳能热水供暖的应用相结合,共同维持着舒适、迷人的内部环境。整座设施比其同等建筑减少了大约60%的能量消耗,并且

每年减少了30.2吨的温室气体（二氧化碳）的排放。
Pam Gaunt考虑周到，引进了艺术照明效果，并从建筑风格和背景及一些本地特有植物中汲取灵感。参照了光合作用原理，每个部件都实时地太阳能发电。主要部件则用来监测和校准这座建筑在不同使用模式下的能量消耗情况。最终的成果在给人们以视觉上的冲击并与人们亲密互动的同时，也潜移默化地实现了教育目的。

A new building in Kings Park is an important commission as this is Perth's most treasured public garden. As part of the Naturescape precinct the Kings Park Environment Awareness Center is a facility and home base for the educational programs run by the Botanic Gardens and Parks Authority. While the building is primarily for school children it is also used for public presentations and functions.

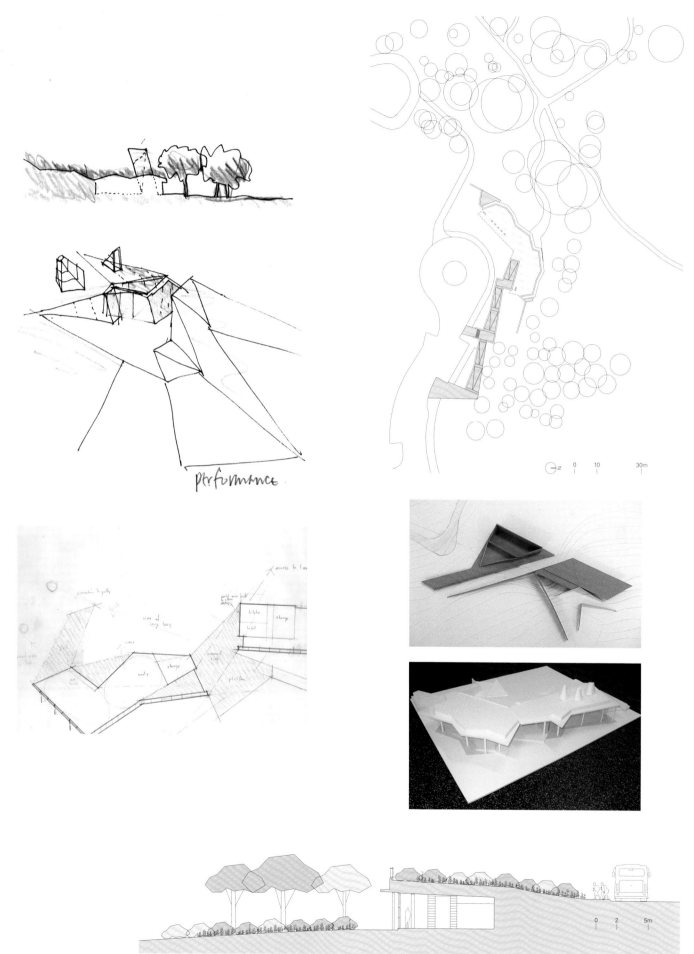

A-A' 剖面图 section A-A'

The expressive and experiential architecture has a significant sustainable agenda focused on generating solar power and reducing energy consumption while participating in a discourse on the aesthetics of sustainability. By folding together the building and its site and re-vegetating the earth roof with natural flora the design prioritises a direct relationship between people, their accommodation and the natural environment. The indigenous flora as main protagonist is reinforced in three maneuvers: the vegetated roofscape; the long entry ramp that descends into the earth bringing visitors to an open terrace submerged in the bushland; and the full-height double-glazed wall that offers users a panoramic view of the setting. The faceted plan, arcing around the site contours, utilizes the benefits of good orientation, and thermal mass plus the adoption of double glazing, LED lighting, recycled materials, cross ventilation and solar water heating are combined to maintain a comfortable and attractive interior. The facility consumes 60% less power than an equivalent building and it reduces greenhouse gas(CO_2-e) emissions by 30.2 tonnes per year.

Pam Gaunt's thoughtfully integrated art use light and draw inspiration from the architecture and its setting and some specific indigenous plants. Each piece in real time is solar-powered, a reference to photosynthesis, and the main piece monitors and calibrates the building's power consumption during its different modes of use. The results are visually engaging, interactive, and subtly didactic works.

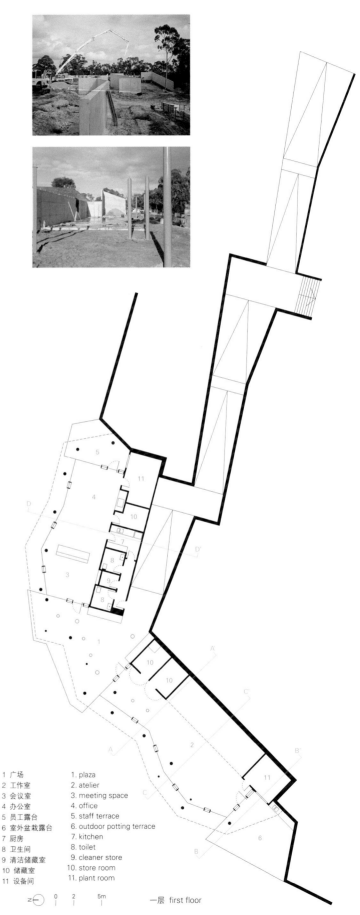

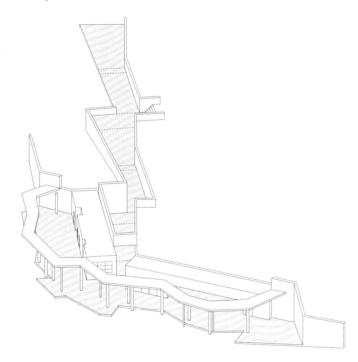

项目名称：Kings Park Environment Awareness Center
地点：Kings Park, Perth, Western Australia
建筑师：Donaldson + Warn
项目负责人：Daniel Grinceri, Steven Postmus
结构工程师：Capital House 机械/电力工程师：Sinclair Knight Merz
水利工程师：Hydraulics Design Australia
景观建筑师：Plan E 室内建筑师：Jennifer Vos
建筑专业研究生：Talya Mossenson, Dean Ismail, Adam Reynolds
助理：Robert Hutchinson
承包商：Pindan 艺术家：Pam Gaunt
甲方：Botanic Gardens and Parks Authority (BGPA)
功能：atelier space, administration area, staff and public amenities
用地面积：7,500m² 有效楼层面积：327m² 平台和斜坡面积：821m²
造价：AUD 2.9m 委托时间：2009.1 竣工时间：2012.4
摄影师：©D-Max Photography(courtesy of the architect) - p.96~97, p.98, p.101, p.102top, p.103, p.104
©Robert Frith, Acorn(courtesy of the architect) - p.102bottom

1 广场	1. plaza
2 工作室	2. atelier
3 会议室	3. meeting space
4 办公室	4. office
5 员工露台	5. staff terrace
6 室外盆栽露台	6. outdoor potting terrace
7 厨房	7. kitchen
8 卫生间	8. toilet
9 清洁储藏室	9. cleaner store
10 储藏室	10. store room
11 设备间	11. plant room

一层 first floor

101

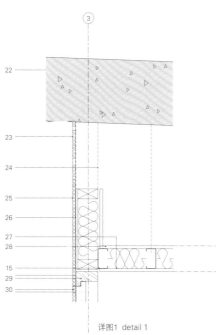

详图1 detail 1

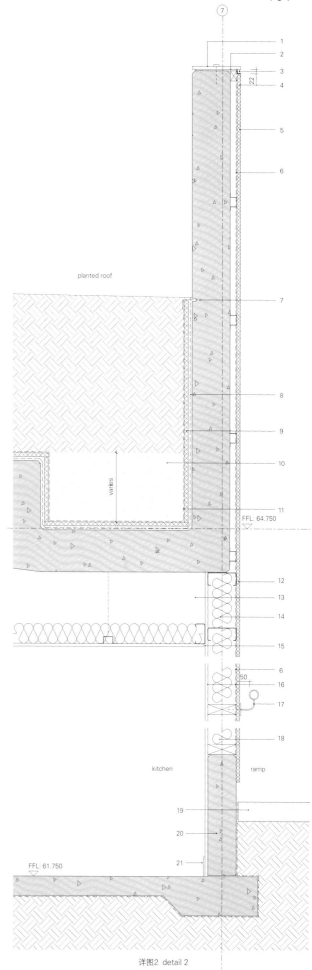

详图2 detail 2

1. 15mm CFC capping
2. mastic bedding
3. silicon seal
4. S.S. flashing
5. CONC. upstand
    OSB cladding fixed to top hats
6. sarking membrane
7. reglet cast into concrete wall
    with flashing over WPM
8. welded water proof membrane
    to floor & walls of concrete deck
9. atlantis drainage system
10. subsoil drainage with 20mm blue metal fill
11. drainage gully to falls formed in CONC. slab
12. sanded and sealed exterior grade 18mm OSB
     lining to studwall
13. suspended flush plasterboard
     ceiling with insulation as SPEC.
14. stud wall with acoustic insulation throughout
15. 10mm shadow detail(TYP)
16. flush plasterboard lining to studwork
17. 50 DIAM. S.S. handrail at 1200 CTRS. as SPEC.
18. stud framed wall to suite thickness
     of retaining wall
19. exposed aggregate CONC. floor to ramp
20. CONC. retaining wall to struct
21. selected ALUM. skirting
22. raking concrete soffit
23. perforated OSB cladding to detail
     for MECH. ventilation
24. line of 100x150 RHS
     column shown dotted beyond.
25. perforated OSB cladding
     as SPEC. on stud framing
26. fabric-faced acoustic insulation as SPEC.
27. flush P. B. ceiling
28. plasterboard lining internal to plenum
29. timber door frame
30. OSB lining concealed
     fixed to solid core
     door with concealed
     hinges to detail

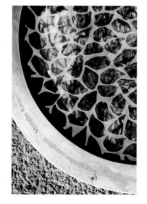 

1. welded capping to CHS post
2. welded end plate to RHS outrigger
3. GALV. 50 SHS rail
4. 10PL galvanized cleat welded to CHS column
5. 80 CHS post GALV. fixed to concrete upstand
6. ALUM. expanded mesh fence
7. GALV. 25x50 RHS outrigger welded to SHS
8. S.S. bolts with epoxy fixings to struct detail/reseal membrane where penetrated
9. waterproof membrane
10. coat CHS post in ground with anti-corrosive paint to struct SPEC.
11. welded base plated to CHS
12. atlantis drainage system
13. 2° fall to CONC.
14. drip groove
15. packer to suit
16. silicone seal
17. ALUM. glazing frame recessed in soffit
18. silicone bead
19. CONC. soffit
20. selected roller blinds fixed between CHS columns
21. thermotech double glazing system
22. S.S point fixed IFU's spaced to manufacturers recommendations
23. 100 CHS COL. beyond paint finish
24. selected marmoleum floor
25. thermoplastic spacer
26. struct silicone seal
27. ALUM. glazing frame recessed in concrete slab with sub-sill
28. CONC. kerb to crushed compact Gingin quartz skirting
29. water proof membrane to underside of sill
30. acoustic underlining to hood of MECH.

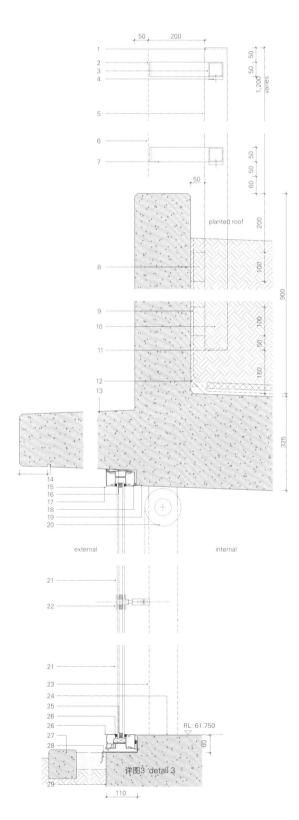

详图3 detail 3

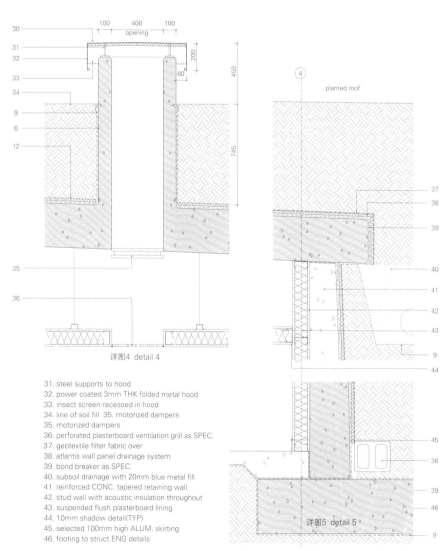

详图4 detail 4

31. steel supports to hood
32. power coated 3mm THK folded metal hood
33. insect screen recessed in hood
34. line of soil fill 35. motorized dampers
35. motorized dampers
36. perforated plasterboard ventilation grill as SPEC.
37. geotextile filter fabric over
38. atlantis wall panel drainage system
39. bond breaker as SPEC.
40. subsoil drainage with 20mm blue metal fill
41. reinforced CONC. tapered retaining wall
42. stud wall with acoustic insulation throughout
43. suspended flush plasterboard lining
44. 10mm shadow detail(TYP)
45. selected 100mm high ALUM. skirting
46. footing to struct ENG details

详图5 detail 5

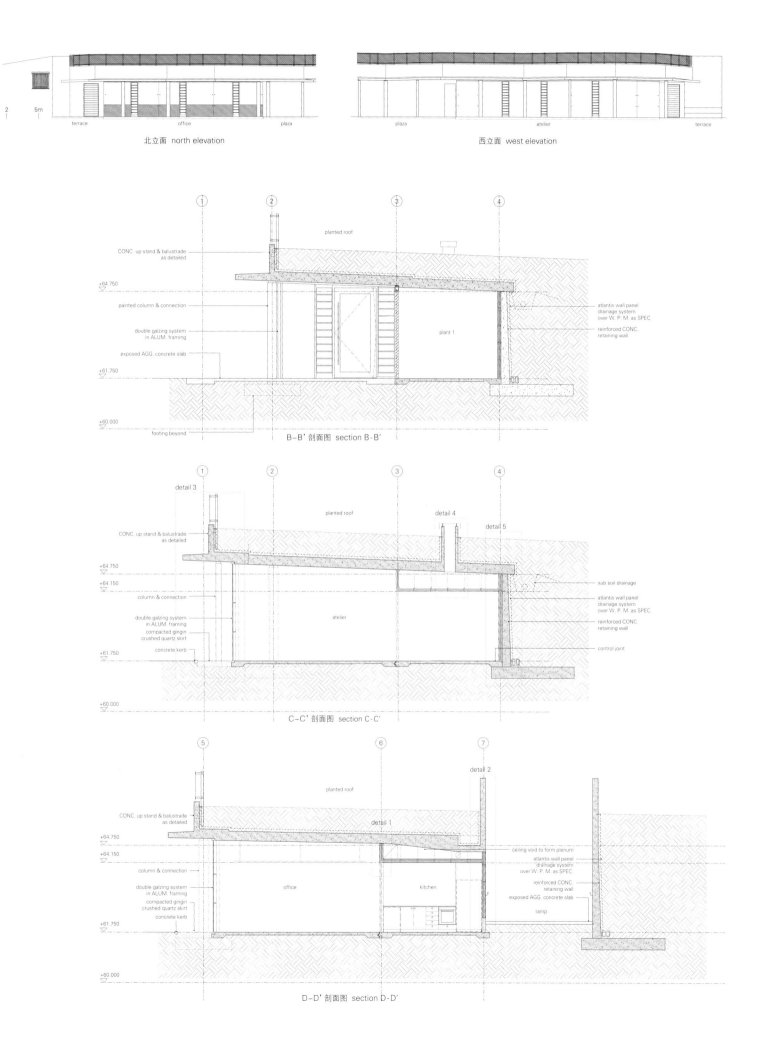

儿童空间 Kidspace

Lucie Aubrac学校
Laurens & Loustau Architectes

位于法国图卢兹的Lucie Aubrac学校项目是由法国Laurens&Loustau建筑事务所设计的。这座设计紧凑的综合教育设施包括一座覆盖整个场地的托儿所。

小学是这座建筑的基础工程,其体量纤细直立,地面层高。带有圆孔的屋顶连接着两边校园,为孩子们提供保护,也为教师提供了一个观察孩子们的有利位置。

树木在人行道上投下美丽的阴影,学生们在繁茂的绿植的包围下尽情嬉戏。

项目设计俏皮地参照了一盒彩色铅笔的形象,整个体量被一层不连续的木质表皮覆盖,可以阻隔日晒,同时也为覆顶的操场提供过滤保护。

Lucie Aubrac School

Designed by the French architects Laurens & Loustau, the Lucie Aubrac School is located in Toulouse, France. This compact educational complex consists of the nursery which covers the entire plot.

The base's project, the primary school, is a thin and vertical volume with high ground floor, with a punched roof between connecting both of the schoolyards, protecting the children and giving onlookers a vantage point from which to observe them.

The pupils enjoy this leafy area, where the trees cast beautiful shadows on the pavement.

In a playful reference to a box of colored pencils, the volume is covered by a discontinuous skin of wood which offers protection from the sun and a filter for the covered playground.

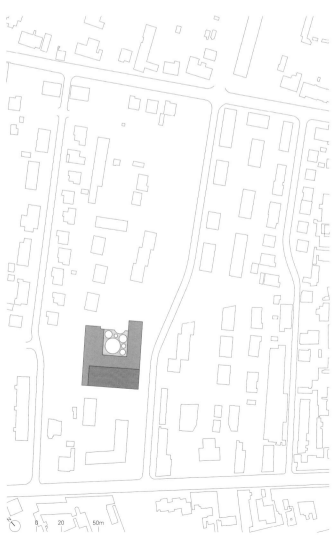

detail 6　东南立面 south-east elevation

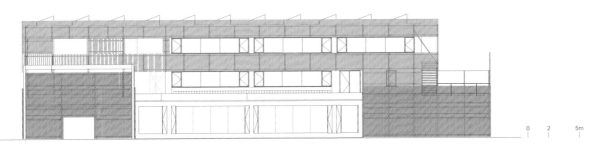

东北立面 north-east elevation

0　2　5m

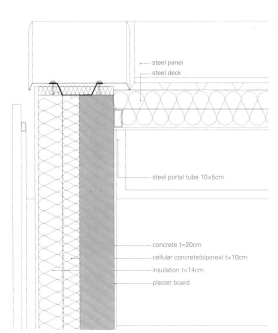

详图1 detail 1

- wood pencil
- steel frame
- flat
- upright
- coat t=20cm
- insulation t=30cm
- cellular concrete t=20cm

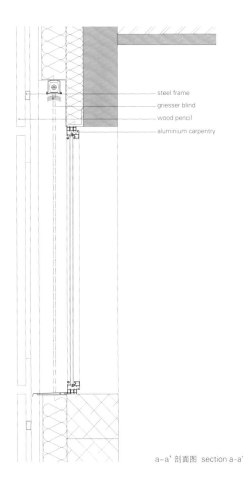

- steel panel
- steel deck
- steel portal tube 10x5cm
- concrete t=20cm
- cellular concrete(siporex) t=10cm
- insulation t=14cm
- plaster board
- steel frame
- griesser blind
- wood pencil
- aluminium carpentry

a-a' 剖面图 section a-a'

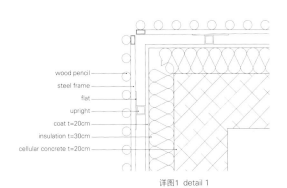

2500 m² public garden

pond

2495 m² Bldg area

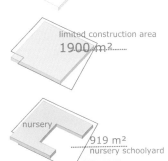

limited construction area 1900 m²

nursery 919 m² nursery schoolyard

elementary school 1167 m² elementary schoolyard

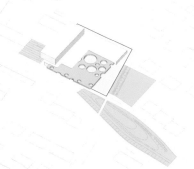

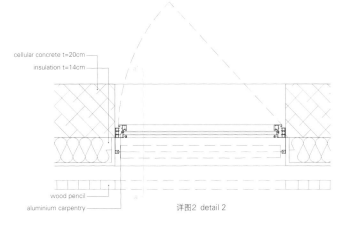

- cellular concrete t=20cm
- insulation t=14cm
- wood pencil
- aluminium carpentry

详图2 detail 2

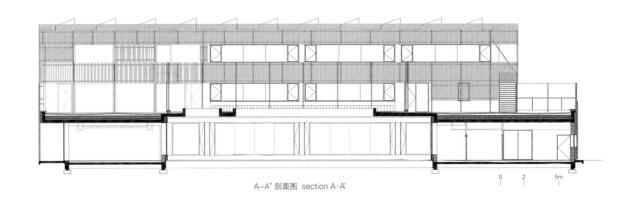

A-A'剖面图 section A-A'

B-B'剖面图 section B-B'

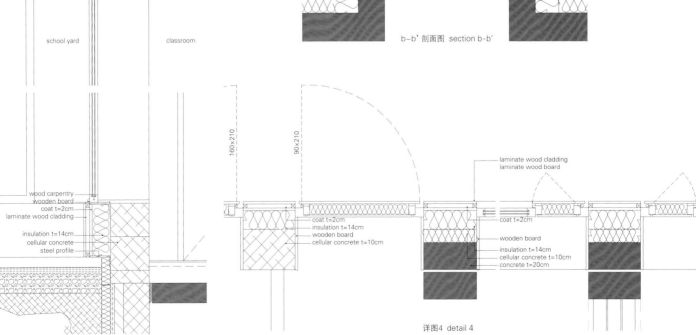

详图3 detail 3

b-b'剖面图 section b-b'

详图4 detail 4

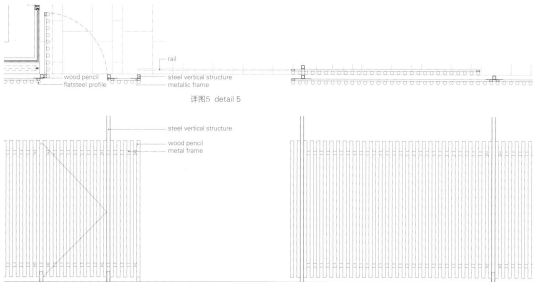

详图5 detail 5

详图6 detail 6

1 初级班
2 教室
3 餐厅
4 增建的餐厅
5 幼儿班
6 办公室

1. elementary class
2. classroom
3. restaurant
4. additional restaurant
5. nursery class
6. office

三层 third floor

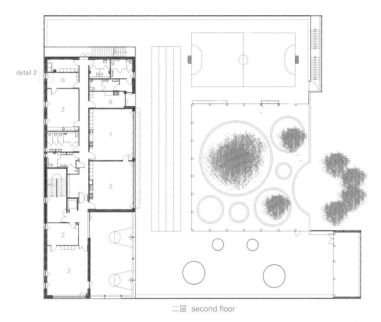

二层 second floor

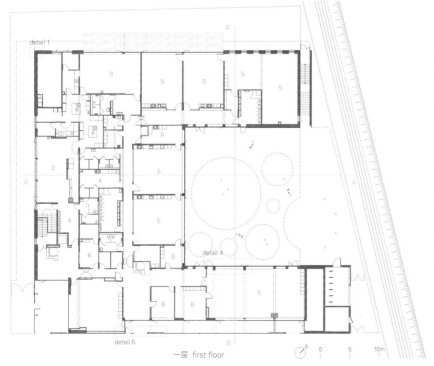

一层 first floor

114

项目名称：Lucie Aubrac School
地点：Toulouse, France
建筑师：Laurens & Loustau Architectes
项目建筑师：Marc Laurens, Pierre Loustau
设计团队：Joanne Pouzenc, Laurent Didier
项目经理：OTCE Organisation
施工单位：GBMP
承包商：Ville de Toulouse
总建筑面积：2,570m² 有效楼层面积：2,495m²
设计时间：2009 施工时间：2012 竣工时间：2013
摄影师：©Stéphane Chalmeau(courtesy of the architect)

Ama'r儿童文化馆

Dorte Mandrup Arkitekter

儿童文化馆提供各种游乐设施和适合所有年龄儿童的活动项目。

这座建筑位于街角,中和了毗邻的高低不同的砖房。但是,在其形式和使用材料上,该中心是相当凸显的。方形窗户穿插在一个银色的铝质表皮间,表皮覆盖在屋顶和外墙上,降低的拐角处可让阳光进入后院。儿童文化馆的表现力是令人惊讶且富有想象力的:屋顶和外立面的处理方式相同,与普通住宅不同,文化馆没有"开端"和"终止"。

室内设计的布局如同一座建在山上的小村庄,提供了丰富的空间体验和迷人的视觉联系。该设计的一个重要贡献是建造了一系列的工作坊,让孩子们为Amager岛的儿童文化中心发挥自己的创意。

这座建筑是在丹麦建筑法对低能量的房子做出规定后建成的,其具有高度保温的外围护结构,低U值(1.1)双层玻璃窗、热回收通风、人工照明、最佳热量分布的加热地板,从而降低温度流动。天窗也减少了对人工照明的需求。Amager,发音为"Ama'r",是丹麦人口最稠密的岛屿,是首都哥本哈根的一部分。

在Amager岛上,Øresundsvej街区多年来一直是丹麦的嘻哈文化中心,废弃的工业建筑早已成为哥本哈根地下音乐的录音室。Øresundsvej已经成为一条要道。这条街,从西部的主街道——Amagerbrogade一直延伸到东部的阿迈厄海滩,特点是破旧的住房和城市空间结合,是一片无人问津的、缺乏活力的街区。

2005年,怀着强化现有文化机构的目标,该地区被选中成为哥本哈根众多综合市区的重建项目之一。尤其对临近街区的Amager Bio音乐厅和儿童文化馆付出了很大的努力。

街区改善项目以现有的文化力量为焦点,因为它使这些项目能够与现有的自然环境和平共处。建筑师的目标不仅是要改善物质条件,而且还要给街区一个有吸引力的特色和身份。

视觉艺术家Kerstin Bergendal,被任命来实现2006年至2007年筹备阶段的对话,Dorte Mandrup建筑事务所也参与其中,利用抽象原则和非理性的幻想影像来工作。这一过程涉及从8岁至14岁的儿童到不同类型职业的成人,包括文化中心的工作人员、图书馆员、教师、音乐家、视觉艺术家和演员的参与,理解儿童世界的成人可以解释他们的活动。

一系列的研讨会得以举行,五场为孩子,五场为成年人。

每次的研讨会通常持续三个小时,然后停办几天去考虑新的想法和意见。

研讨会提出过的主题包括:"太空经历和冒险""理想的玩耍场所""梦想中的住房",重点回顾"有趣的情景"以及"描述对于个体参与者有重要意义的地方"。大人和孩子可以使用的工具包括绘图和建模。随后Kerstin Bergendal将参与者描述自己在讲习班中的收获、想法和作品的情景拍摄了下来。成果的总结概要集合了九大原则,作为空间规划的基础,为建造鼓励孩子通过艺术、音乐和舞蹈来表达自己想法的建筑提供了基础。

Ama'r Children's Culture House

The Children's Culture House offers various play facilities and programmes for children of all ages.

Located on a street corner, the structure mediates between the adjoining brick buildings, which are of different height. In it's form and use of materials, however, the centre is quite distinct. A silvery aluminium skin, perforated by square windows is drawn over the roof and outer walls and the lower height at the corner allows sunlight to enter the rear courtyard. The expression of the Children's Culture House is surprising and imaginative: the roof and facades are treated the same, and the House does not have a "start" and "end" as ordinary houses do.

The interior is organized as a mountain village – providing a number of spatial experiences and fascinating visual links. An important contribution to the design was a series of workshops, allowing the children to develop their own ideas for the Children's Cultural Centre on Amager.

The building is built after Danish Building Regulations for low energy houses. Mainly actions are highly insulated building envelope, doubled-glazed windows with low u-value (1.1), ventilation with heat recovery, artificial lighting, heated floor for optimal heat distribution and thus lower flow temperatures. Skylights reduce the need for artificial lighting.

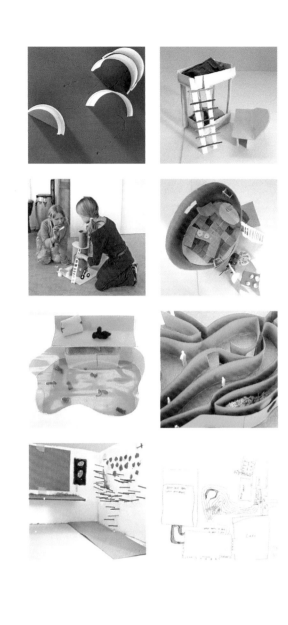

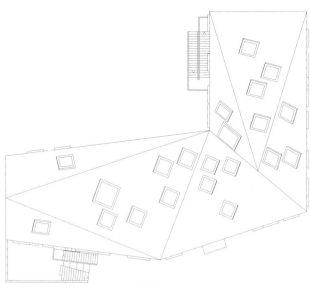

屋顶 roof

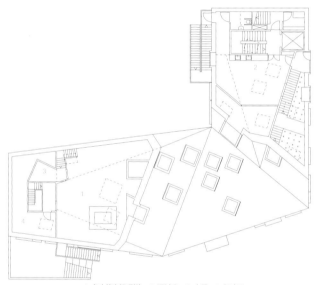

1 多功能空间/剧院 2 研讨室 3 夹层 4 行政区
1. multipurpose space/theater 2. workshops 3. mezzanine 4. administration
三层 third floor

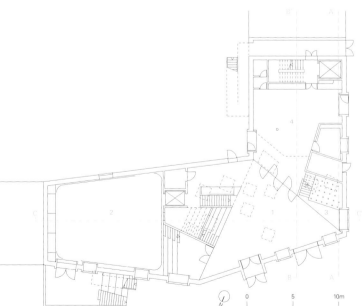

1 广场/门厅 2 多功能空间/剧院 3 "登山"区 4 研讨室
1. plaza/foyer 2. multipurpose space/theater 3. mountain climbing 4. workshops
一层 first floor

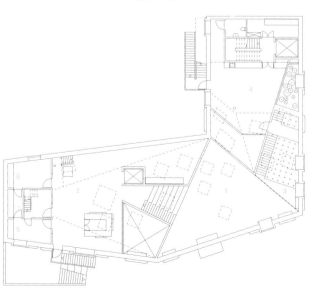

1 广场/门厅 2 咖啡室 3 "登山"区 4 舞蹈工作室 5 行政区
1. plaza/foyer 2. cafe 3. mountain climbing 4. dance studio 5. administration
二层 second floor

Amager, pronounced "Ama'r", is Denmark's most densely populated island, an island which is part of the capital, Copenhagen. On Amager the neighbourhood around Øresundsvej has been a center for Danish hip hop culture for many years, and the derelict industrial buildings have long housed recording studios for Copenhagen's underground music scene. Øresundsvej has been quite a thoroughfare. The street, which runs from the main street Amagerbrogade in the west to Amager Beach in the east, was characterized by dilapidated housing and urban spaces – a neglected and uninspiring neighbourhood.

In 2005, the area was selected to become one of Copenhagen's numerous integrated Urban Renewal projects, with the ambition of strengthening the existing cultural institutions. Efforts were among others, and made concerning the neighboring Amager Bio Music Hall and the Children's Cultural House.

The neighborhood improvement project took the existing cultural strength as its focus, as it allowed them to work with the existing physical environment. The goal was not only to improve physical conditions, but to give the neighborhood an attractive character and identity.

Visual artist, Kerstin Bergendal, was appointed to realize the preparatory phase of dialogues in 2006-2007, in which Dorte Mandrup Arkitekter participated, working with abstract principles and irrational dream images. The process included both children from the ages 8-14 and adults with different kind of professions: members of the staff of the cultural center, librarians, teachers, musicians, visual artists and actors. Adults who understand the world of children can interpret their activities.

A series of workshops was held – five with children and then five with adults.

The workshops usually lasted for three hours at a time and then had a break for a few days to have time to consider thoughts and ideas.

Some of the themes raised was "experiences and adventures with space", "the ideal place to play", "dream house", focusing on "recalling interesting situations" and "describing places with importance to the individual participant". The tools being both drawing and modeling are available for both children and adults. Subsequently Kerstin Bergendal filmed the participants describing their input, ideas and their work during the workshops. A summary of the outcome formed a collection of nine principles, which served as a basis for the spatial program, which is a basis for a building that encourages children in expressing themselves through art, music and dance.

A-A' 剖面图 section A-A'   B-B' 剖面图 section B-B'

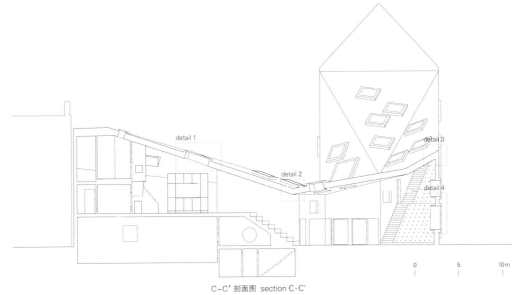

C-C' 剖面图 section C-C'

详图1 detail 1

详图2 detail 2

详图3 detail 3

详图4 detail 4

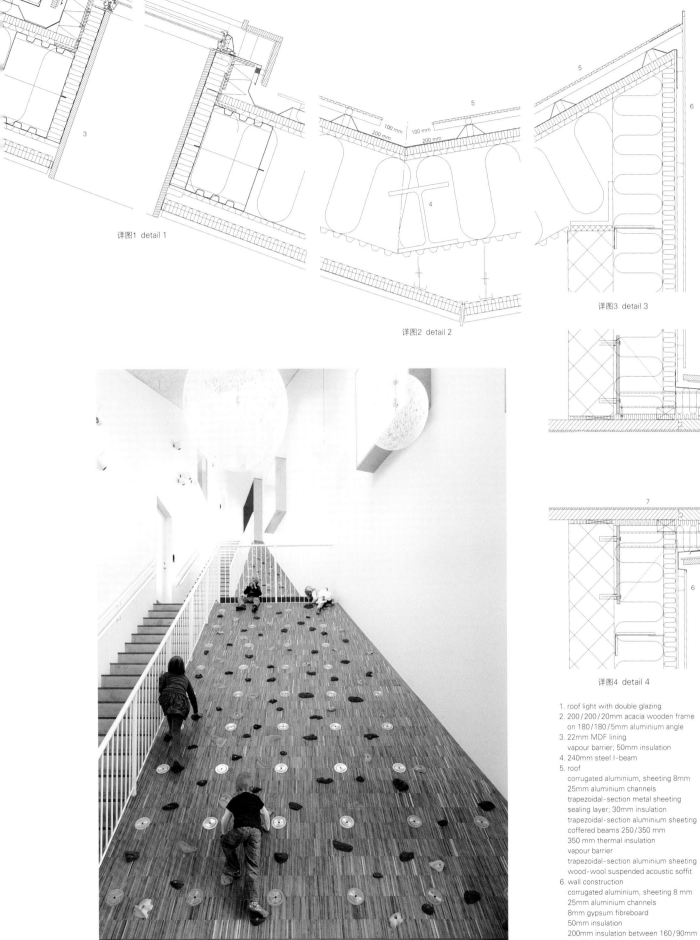

1. roof light with double glazing
2. 200/200/20mm acacia wooden frame on 180/180/5mm aluminium angle
3. 22mm MDF lining
   vapour barrier; 50mm insulation
4. 240mm steel I-beam
5. roof
   corrugated aluminium, sheeting 8mm
   25mm aluminium channels
   trapezoidal-section metal sheeting
   sealing layer; 30mm insulation
   trapezoidal-section aluminium sheeting
   coffered beams 250/350 mm
   350 mm thermal insulation
   vapour barrier
   trapezoidal-section aluminium sheeting
   wood-wool suspended acoustic soffit
6. wall construction
   corrugated aluminium, sheeting 8 mm
   25mm aluminium channels
   8mm gypsum fibreboard
   50mm insulation
   200mm insulation between 160/90mm and 60/40 mm steel angles
   180mm precast concrete element
   windowsill
7. Pro Tec classic in aluminum and wood
   the indoor frame is 40mm two-layer birch plywood

项目名称：Children's Culture House Ama'r
地点：Øresundsvej 8B, 2300 København S, Denmark
建筑师：Dorte Mandrup Arkitekter
施工：Nøhr & Sigsgaard Architects
设计经理：Dorte Mandrup Arkitekter
工程师：Dominia
承包商：Anker Hansen & co.
甲方：City of Copenhagen
用地面积：1,085m²
施工时间：2011.10—2012.11
竣工时间：2013.2
摄影师：
©Bo Bolther(courtesy of the architect)-p.124, p.126 bottom-left, p.126 top-left
©Jens Lindhe(courtesy of the architect)-p.120, p.126 top-right
©Torben Eskerod(courtesy of the architect)-p.118~119, p.122, p.123, p.125, p.126 top-left, p.127 bottom-left, top-right, bottom-right

Lasalle Franciscanas学校高架运动场

Guzmán de Yarza Blache

这一委托任务起源于由于大量的学生和家长通常在白天聚会，因此校方要增加庭院面积的要求。这些学生和家长会影响本应在此进行的体育运动和休闲活动。

这个庭院宽33m，长35m，为东南朝向。它由现在的U形且带有两翼的学校形成，其中一翼形成于20世纪50年代，而另一翼形成于20世纪70年代。

身为学校的一个事实就是，建筑师必须在夏季几个月中完成这项建筑工程。这一事实要求建筑师立刻考虑建造一个可以在几天里就能够建成的预制混凝土结构，并且它也能够覆盖距离地面13m的体量。

该项目的另一个关键因素就是必须清除院子里现有的两颗树，并且种植新结构中的植物。为此，建筑师设计了70m长的考顿钢花盆，里面种植了300多种常春藤植物，在几年内，这些植物将覆盖整个金属圆顶。

金属圆顶由双层镀锌钢形成，因此，其中一层可以帮助常春藤作物生长，而另一层能够耐受年轻人进行的球类相关运动所带来的冲击。

地面拥有一个有机形状的花园式长椅，内设不同的植物种类。学生和家长可以坐在上面并且观察它们。

新建筑与学校其他部分的关系也必须得到解决，因此，一个45m长的斜坡连接了地面层与中间层以及高架球场。

另一个有机斜坡也在其中，它可以让孩子们从婴儿区到达地面层（部分在球场下方）的庭院里。

之后对学校的访问证明了项目的成功，且它被学生快速且标志性地认同，学生们亲切地称它为"鲸鱼"。

Lasalle Franciscanas School Elevated Sports Court

The commission is originated by the need from the school to augment the total surface of the courtyard that due to the great amount of students and parents that usually gather during the day, can sometimes obstruct the correct developing of the sports and leisure activities that should take place in it.

The courtyard is 33 meters wide per 35 meters long and has a south-east orientation. It is formed by the existing school that has a U form with two wings, one from the 1950s and another one from the 1970s.

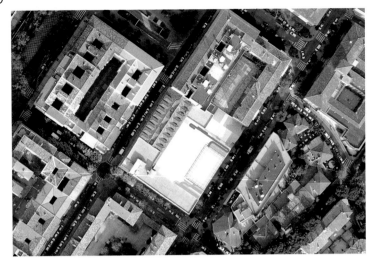

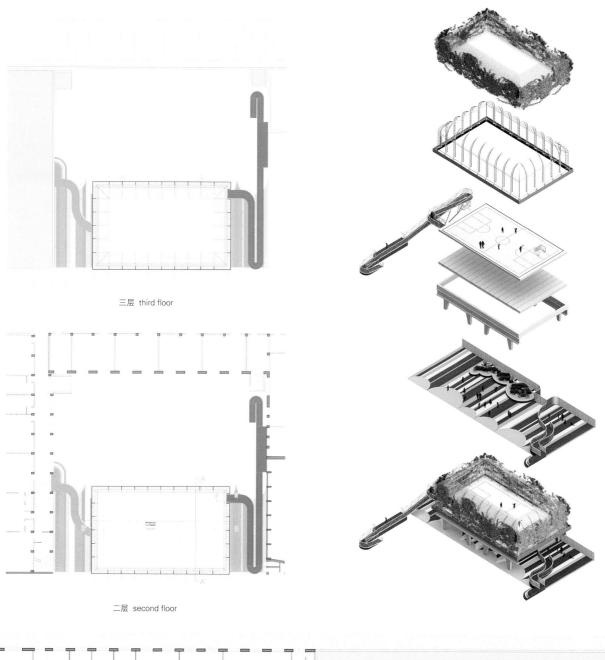

三层 third floor

二层 second floor

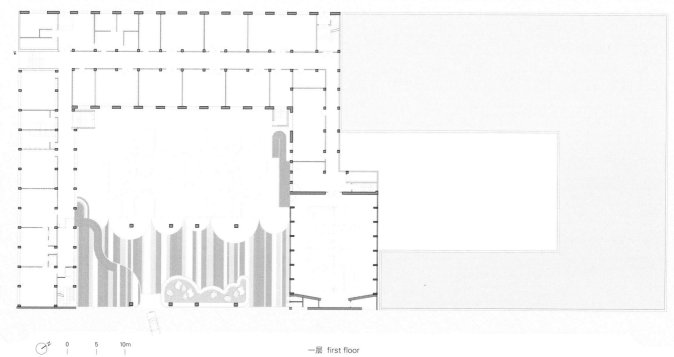

一层 first floor

项目名称: Elevated Sports Court at Lasalle Franciscanas School
地点: Calle andrés piquer 5, Zaragoza, Spain
建筑师: Guzmán de Yarza Blache
合作建筑师: Ana Guzmán Malpica, Julien Luengo-Gómez
建筑公司: GM Empresa Constructora
工料测量: Jose Manuel Arguedas
结构工程师: Josep Agustí de Ciurana, PRAINSA
甲方: Lasalle Franciscanas School
建成面积: 350m²
造价: EUR 290,000
竣工时间: 2012.9
摄影师: ©Miguel de Guzmán (courtesy of the architect)

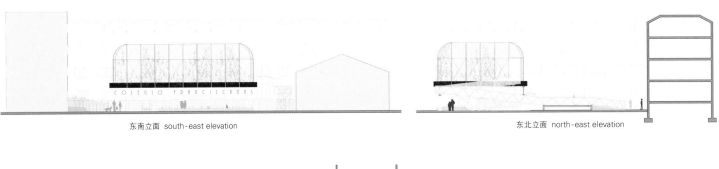

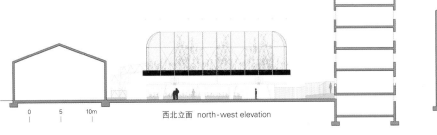

东南立面 south-east elevation

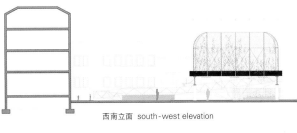

东北立面 north-east elevation

西北立面 north-west elevation

西南立面 south-west elevation

1. steel slab e=10mm
2. simple torsion mesh anchored to the structure
3. compression layer e=10cm
4. concrete slope
5. geotextile sheet
6. layer of compacted gravel e=20cm
7. armed concrete slab
8. synthetic topcoat. based on acrylic resin and synthetic end grain
9. alveolar plate
10. metalic profile L 40.40.10
11. waterproofing membrane
12. structural anchor plate
13. plumbing pipe ø160mm
14. manifold ø150mm
15. galvanized metal profile
16. prefabricated beam
17. prefabricated beam by PRAINSA
18. linear fluorescent luminaire
19. methacrylate board 120x90 cm
20. anchor plate for the basket
21. prefabricated pillars by PRAINSA
22. galvanized steel handrail ø60mms
23. omega metalic profile 120.50.5mm
24. drainage pipe ø150mm
25. tramex grid for water evacuation
26. semi-polished compression layer e=10cm

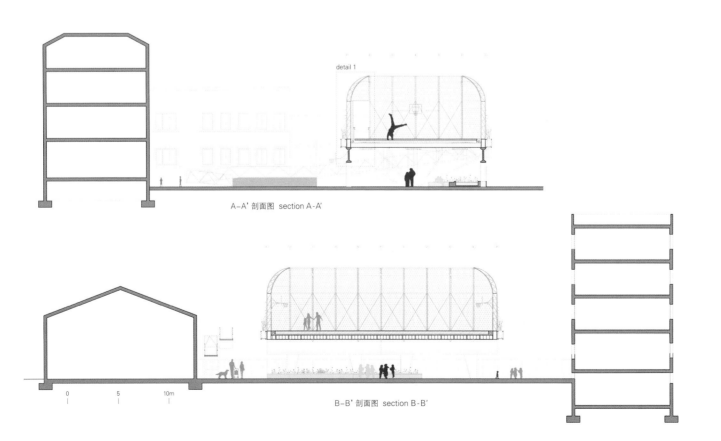

A-A' 剖面图 section A-A'

B-B' 剖面图 section B-B'

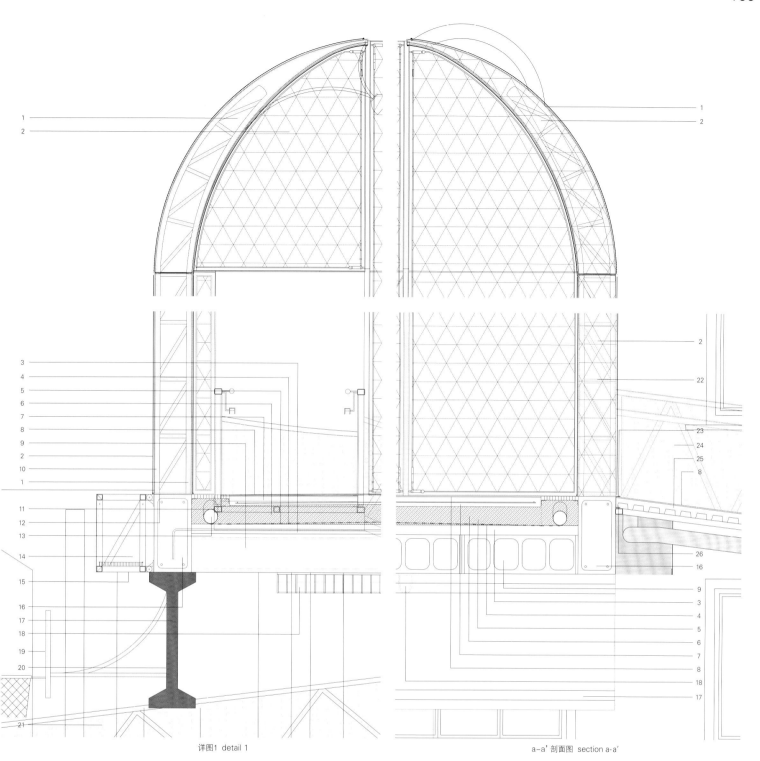

详图1 detail 1   a-a' 剖面图 section a-a'

The fact of being a school meant that the architects had to accomplish the building works exclusively during the summer months. That fact made the architects immediately think about a prefabricated concrete structure that could be built in a couple of days, and that could also solve the 13 meters distance that the architects wanted to cover in the ground level.

The necessary elimination of the two existing trees in the courtyard gave another of the key drivers of the project; the inclusion of vegetation in the new structure. To do so the architects have designed a 70 meters long corten steel flower pot from which almost three hundreds of ivy plants grow, that in a few years will have covered the whole metallic bubble.

That metallic bubble is formed with a double layer of galvanized steel, so one of the layers can help the ivy grow while the other one can resist the practice of teenagers' ball-related sports.

The ground level hosts a garden-bench with an organic shape that includes different species of plants and allows the parents and the students to sit down and observe.

The relation of the new volume with the rest of the school also had to be solved, for which a soft 45-meter ramp was designed to connect the ground level with an intermediate level and the elevated court.

Another organic ramp was also included to let the children from the infantile area get out to their courtyard's zone, also in the ground level and partly under the court.

The later visits to the school have revealed the success of the project and its fast iconic assimilation by the students, who have kindly called it "The Whale".

Seung, H-Sang

测试建筑的力量
Testing the Strength of Architecture

Hyungmin Pai + Seung, H-Sang

Hyungmin Pai：去年冬天，我参观了GunUi SaYa公园的MoHeon、WooJeong和HyunAm建筑。我原以为，这些一定是过去十年中最难建的项目。因为周围的大自然环境似乎是主导力量。在GunUi SaYa公园，自然是主导，MoHeon建筑位于一座非常迷人的日式花园。决定在这些小建筑中呈现什么样的建筑绝对不是一件容易的事情。我很好奇这些项目是否不同于您的其他作品呢？

承孝相：未必，但如果你确实感到有区别，这可能是因为我对建筑的兴趣已经转移。上述项目的规模较小。GunUi SaYa公园占地面积约为100 000m²，但在视觉上看起来较大，而且在视线范围内没有其他人造项目。这些小结构应该如何出现在这种绝对的自然环境中呢？所面临的困难就在于这个问题中，这制约了整个设计过程。亦由于这，在寻找"读风景"的基础，即在寻找这片土地上的自然和生活印记的基础方面存在难度。

Pai：SaYaWon是迷人的花园。你们一定关注过可融入这样环境的建筑特点。

承孝相：这是当然。从一开始，花园就是焦点，而不是建筑物。最初的计划是建造一个日式茶馆，以相称一个日式庭院。所有的详细资料，包括位置，已经由日本景观建筑师决定。

但是，在花园竣工时，客户改变了主意，要求我的事务所负责建设。这正是我所希望的。即使原计划通过了，也将不太可能在韩国的土地上幸存下来。我的设计理解优先考虑花园。花园的风景和内容随时间变化。我的独特理念是，建筑应同样与园林保持和谐，且其占地面积为13m²。

Pai：在MoHeon、HyunAm和平度市住宅文化馆，YoungSun Jeong夫人的景观设计起着重要的作用。尤其是在MoHeon，景观是如何与这一进

Hyungmin Pai Last winter, I visited MoHeon, WooJeong, and HyunAm of GunUi SaYa Park. I thought that these must have been the most difficult projects of the last 10 years. It was because the surrounding nature seemed like such a dominant force. The nature in GunUi SaYa Park is overwhelming and MoHeon is located in a very glamorous Japanese garden. It must not have been easy to decide what kind of architecture should be present in these small buildings. I am curious whether these projects were different from your other work.

Seung, H-Sang Not necessarily, but if you did feel a difference, it would be because my interest in architecture had shifted. The above projects were all small in scale. The GunUi SaYa Park site is around 100,000m², but visually looks larger, and there is no other man-made work within sight. How should, and could these small structures appear within such absolute nature? The difficulty was in this question, which governed the entire design process. Also attributable to this was the difficulty in finding a basis for the "Landscript", the inscription of nature and life on the land.

Pai The protagonist of SaYaWon is the glamorous garden. There must have been some concerns about the character of the architecture that would enter such an environment.

Seung Of course. From the start, the garden was the focus, rather than the architecture. The original plan was to build a Japanese tea house befitting a Japanese garden. All the detailed particulars, including location, were already decided by the Japanese landscape architect.

But, around the time when the garden was completed, the client changed his mind and asked for Seung, H-Sang architecture. That's what I hoped for. Even if the original plan had gone through, it would have been unlikely to survive on Korean land. I designed with the understanding that the garden takes precedence. The scenery and content of the garden change over time. The singular notion was that the architecture would do the same in harmony with the garden, especially given its size of 13m².

Pai In MoHeon, HyunAm, and Pingdu Housing Culture Center, Mrs. YoungSun Jeong's landscaping plays an important role. Especially in MoHeon, how was landscaping involved in the process?

Seung Given the small space of the land, the decision was made

HyunAm, 一座黑色的小屋_HyunAm, A Black Cottage
申东烨文学博物馆_Shin DongYeop Literary Museum
平度市住宅文化馆_Pingdu Housing Culture Center
测试建筑的力量_Testing the Strength of Architecture/Hyungmin Pai + Seung, H-Sang

程相关联的呢？

承孝相：考虑到土地的空间很小，我们决定设计空间，来创造一座房屋的形式。由于客户原本希望是韩式花园，我对外部空间设计的参与性也最小化了。从一开始，YoungSun Jeong夫人是我脑海中完成这个任务的人选。在这个过程中，我的素描和切片到处都标有YoungSun Jeong的名字，这就意味着标记的区域完全托付给她。然而，当我参观接近完工的MoHeon时，花园已成为一片茂密的竹林。这是一个令人印象深刻的景象，但它不是我曾设想的Jeong式景观。我说服了客户，他同意重建，把它变成一座真正属于Jeong的花园。有一年春天，我被召唤到场地，在那里我很满意看到的新景观。

Pai：包括Jeong的花园在内，MoHeon至少包含四个空间层。韩国传统房间和餐厅是可居住的空间，而它们之间的池塘空间是一处非居住

的、空置的空间。因此，人们可以把这处空置的空间作为中心，但它不能放置任何实物。根据四季变化和每天时刻的变化，照明、发生在每个房间的事件、空间的居住者之间的关系也随时改变着。同样，消息的内容和价值通过建筑也在改变。当有人在韩式房间里唱歌时，房间就成了一个富有魅力的舞台；餐厅变观众厅。如果屏风对韩式房间进行了围合，餐厅就会成为沉思的空间，位于有水和树木的花园之内。

承孝相：正是，MoHeon处在一块接近于方形的场地，场地的一侧是现有的建筑，其他三个侧面是由独栋住宅和公寓相结合的街区。因此，狭小的空间被划分成若干层，使每处独特的空间看起来较大，相对时尚。此外，每处空间被赋予了不同的个性来营造浓郁的氛围。场地基本上被分为五个平行层次：前院、餐厅、中央院子、韩国传统房间和后院。人们的视野可以从韩式房间穿过餐厅，并延伸至前院、黑色考顿钢墙，甚至以外。

to design the space in order to create the form of a house. Since the client originally wished for a Korean garden, my participation in the outside space was minimized. From the start, Mrs. Young-Sun Jeong was in my mind for this task. During the process, my sketches and sections were everywhere labeled with the name, "YoungSun Jeong." This meant that the labeled area was fully entrusted to her. However, when I visited MoHeon nearing its completion, the garden had become a thick bamboo grove. It was an impressive sight, but it was not the Jeong landscape that I had envisioned. I persuaded the client and he agreed to rebuild it into a garden true to Jeong's vision. One spring, I was called to the site, where I was satisfied to see the new landscape in place.

Pai Including Jeong's garden, MoHeon contains at least 4 spatial layers. The traditional Korean room and the dining room are inhabitable spaces, whereas the pond space between them is a non-inhabitable, empty space. Thus, though one can place one's heart in that space, it cannot be physically occupied. According to the changing seasons and the hours of the day, the lighting, the events that take place in each room, and the relationship between the occupants of the space change. Likewise, the story and value of the message relayed by the architecture also change. When someone is singing in the hanOk room, it becomes a glamorous stage; the dining room becomes an audience hall. If the screens seal off the hanOk room, the dining room becomes a contemplative space amongst a garden of water and trees.

Seung That's right. The nearly square site in which MoHeon is placed is contained on one side by the existing building, and the other three sides by surrounding single house and apartment complexes of the neighborhood. Hence, the small space was divided into several layers so that each unique space appears larger in relative fashion. Furthermore, each space was given a different personality to create a rich atmosphere. Fundamental to the location are five parallel layer spaces: front yard, dining room, middle court, traditional Korean room, and back yard. The perspective from the Korean room penetrates the dining room and extends to the front yard, black corten steel wall and beyond.

Pai In that perspective, Jeong's garden space is crucial in establishing the spatial relationships.

MoHeon的花园，YoungSun Jeong设计
MoHeon's garden, designed by YoungSun Jeong

Pai：从这个角度看，在Jeong的花园空间中，建立空间关系是至关重要的。

承孝相：是。只有她本来有可能完成，把这个任务委托给其他任何人都是很难完成的。

Pai：在GunUi SaYa公园的实例中，首先是要找到一个合适的地点。请和我们分享一下这个过程。

承孝相：场地位于100,000m²土地的中心，在环境风水方面，这个区域是核心。在这个地方，看不到任何人造结构，只有山的轮廓所呈现的层次。在这里，人们可以感受到绝对的孤独。客户无意在这里建造一个大型结构；相反，一个小公馆，将成为未来建造一座植物园的开始。因此，我想通过这个结构来象征植物园的个性。同样MoHeon和HyunAm是我为这个住所提议的名字。

近来，YoungSun Jeong为GunUi SaYa设计了草图，并决定它不会对公众完全开放；相反，将设置一个预约系统，只能使一定数量的人们享受它的孤独感。

Pai：HyunAm的规划是简单且特点突出的。当你进入大楼，大楼会消失在自然的框架中。有广阔的大自然为背景，由Jeong设计的屋顶景观彰显着建筑的意志。有趣的是，屋顶的景观设计使入口区的感觉比较做作。虽然主题与MoHeon的主题类似，但是作为私人住宅，HyunAm散发出不同的感觉，即更精致的景观。在HyunAm中，你是如何和Jeong合作的？

承孝相：客户的家庭墓地就在GuMi附近。场地有七个坟墓，三代人都安葬在此墓地，这里最近因开发压力而面临着一些困难。客户对死亡有着独到的见解。他不希望建成一个墓地，甚至是阴森的停尸处。然而，

Seung Yes. Only she could have made it possible. It would have been difficult to entrust this task to anybody else.

Pai In the case of GunUi SaYa Park, the start was to find a suitable site. Tell me about that process.

Seung The site is at the center of the 100,000m² land; In terms of geomancy, this area is the core. At this spot, no artificial structure is visible – only the layers of the mountain's silhouette. Here one can feel absolute solitude. The client had no intention to build an immense structure here; rather, a small residence that would be the beginning of an arboretum to be built in the future. Accordingly, I wanted to symbolize the personality of the arboretum through this structure. Likewise MoHeon and HyunAm are names I proposed for this residence.
Recently, YoungSun Jeong drew a rough sketch of the GunUi SaYa Park. It was decided that it would not be totally open to the public; rather, a reservation system would be set up so that a limited amount of people would enjoy it in solitude.

Pai HyunAm's plan is simple and strong in character. When you enter the building, the building vanishes into a frame for nature. With the vast nature as background, it is the rooftop landscape designed by Jeong that reveals the architectural will. It is interesting that the rooftop landscape causes the entry area to feel more artificial. Although the theme is similar to that of MoHeon, as a private house, HyunAm gives off a different feel of a more refined landscape. In which way did you collaborate with Jeong on HyunAm?

Seung The client's family gravesite is nearby GuMi. Seven graves and three generations are in the gravesite, which has recently faced difficulty due to development pressures. The client has a unique perspective on death. He has no wish for a burial or even a charnel house. However, the formal ritual of death chosen by his ancestors must be respected. He decided to provide his ancestors a safer location at GunUi SaYa Park. Thus, the Park will become a land in which life and death coexist. A few artificial facilities exist on the way to enter HyunAm. Five small spaces made of corten steel are dedicated to cherishing and celebrating ancestors. Beginning from such spaces, HyunAm is a house that sprouts from the ground. The logic of the land is artificial and guided by instinct. Once reaching the location, the polarizing scenery faced

位于GunUi SaYa公园内的五处小空间中的一处专用于缅怀和赞美客户的祖先
one of the five small spaces in GunUi SaYa Park, dedicated to cherishing and celebrating the client's ancestors

他的祖先选择的正式的葬礼仪式必须得到尊重。因此他决定给他的祖先在GunUi SaYa公园提供一个安全的场地。因此，该公园将成为生命和死亡并存的土地。进入HyunAm的途中设有一些人工设施。由考顿钢制成的五处小空间专门用于缅怀和赞美其祖先。从这些空间方向看，HyunAm就是一栋从地面拔起的房子。土地的建造逻辑是人为的，由人的本能所引导。一旦到达位置，人们面对的将是完全自然的极致风景。这所房子在这片土地上的存在没有什么意义，主要目的是解决被踩踏的土地。我有信心，YoungSun Jeong可以提供一个解决这个问题的办法。而且，建筑无疑成为土地的一部分。

Pai：尽管是一座历史文化名城，平度却是一座荒凉的城市，几乎没有留下一丝过去的痕迹。相比之下，平度市住宅文化馆的所在位置围绕着茂密的、优美的法国梧桐树而建。在这样的场地情况下，我相信你的建筑策略和Jeong的地面景观形成了独特的关系。

承孝相：该区域的确长满了茂密的古树，任何人看到这片土地都会本能地知道，没有一棵树应该砍掉。我也不得不避开树木，将项目建在裸露的土地上。因此，布局和雕塑的形式已经预先确定。由于没有空间种植额外的树木，它最初似乎没有景观工作要做。但是，Jeong的作用即便是在这种情况下也很重要。通过操纵地面板，她创造了一处带有完全不同情感的空间。现在树木是主题，建筑是背景。在新的入口处，地面层被提高，以纪念原来的位置。这是唯一的解决方案，让平度被人们所记住。

Pai：平度市住宅文化馆的平面显示了与光州双年展附属建筑物相似风格的方案。然而，这两个结构的意图和策略有很大的不同。光州双年展建筑侧重于明确界定的方形外形，而平度市住宅文化馆的结构隐藏在

with absolute nature unfolds. This house came to existence by boring through the ground, and the main concern was to resolve the torn land. I had faith that YoungSun Jeong could provide a solution to this issue. Again, architecture undoubtedly became one with the land.

Pai Despite being a historical city, Pingdu is a desolate city with nearly no trace of the past. In contrast, the site of the Pingdu Housing Culture Center is dense with beautiful platanus trees. In the context of such a site, I believe that your architectural strategy and Jeong's ground landscaping form a unique relationship.

Seung The site was indeed dense with old trees. Anyone who sees this land will instinctively know that not a single tree should be cut down. I also had to avoid the trees and build on a bare patch of land. Thus, the placement and sculpture form were already pre-determined. Since there was no space for planting additional trees, it initially seemed that there was no landscaping work to be done. However, Jeong's role was important even in this setting. By manipulating the ground boards, she created a completely different mood for the space. Now the trees were the subjects and the architecture was the background. At the new entrance, the ground level was raised to commemorate the old location. This was the sole solution to allow Pingdu, a city bare of its historical vestige, to be remembered.

Pai The plan of the Pingdu Housing Culture Center displays a similar motive to the annex building of GwangJu Biennale. However, the intention and strategy of the two structures are very different. In the case of the GwangJu Biennale building, the architecture focuses on clearly defining the square. The structure of the Pingdu Housing Culture Center hides into the background. The smooth exterior of the Pingdu Housing Culture Center is finished with a similar dark tone shared by basalt and corten. Furthermore, with the exception of the long second floor structure that extends to the roads, the elevation contains no frame elements.

Seung That's right. There was no need to emphasize the existence of architecture here. However, the area bordering the city was adjusted accordingly. It was given a concrete colonnade along with the name, "City Gallery." In fact, several events are taking place at this area.

平度市住宅文化馆入口处的现存的法国梧桐树和混凝土柱廊
existing platanus trees and concrete colonnade at the entrance of Pingdu Housing Culture Center

背景中。平度市住宅文化馆的外部饰面为暗色调,材质为玄武岩和考顿钢。此外,除了二层的结构延伸至道路外,其立面不包含任何框架元素。

承孝相:没错。没有必要在这里强调建筑的存在。而且,城市边缘地区得到了相应调整,还增建了一个混凝土柱廊,命名为"城市展览馆"。事实上,的确有一些活动发生在这个区域。

Pai:在平度市住宅文化馆的长方体空间里,建筑是最明显的。内部空间中各个点的视角都朝向地板,自然会引起人们对外部广场的地面景观的注意。尽管如此,总体思路似乎是设计"森林里的房子"。

承孝相:这不就是建筑存在的最初原因吗?建筑本身可能是特殊的,但背景功能是更根本的。

Pai:"建筑所显露的正是它要隐藏的",或者说"这个地区没有建筑,即是有建筑"。这样的表达在实际的设计过程中有意义吗?

承孝相:肯定有。它是建筑师创造生命的必要形式。一些情况下,最初的理念从开始持续到竣工,但更经常的是,这些在实际的设计过程中不停地改变。

Pai:似乎存在两种极端的建筑设计流程。在一种流程中,项目理念首先建立起来,并且得以落实,随后该设计也相应地发展。而在另一种流程中,各种探索和思考一起产生最终设计。当我谈到韩国现代建筑时,我声明承孝相的流程属于前者。你同意吗?你会如何形容这一过程?

承孝相:这当然符合前者的过程。一名建筑师必须对建筑以及我们的生活方式有着核心理念。在我的项目开始时,我先建立其基本概念的语言。这是因为这种语言的定义会组织随后的思想和想法。

Pai:申东烨文学博物馆有不同的组成类型。此外,流线和功能也大

Pai Architecture is most apparent in the Pingdu Housing Culture Center in the long box space. The perspective of the inner space points toward the floor, naturally drawing attention to the ground landscape of the exterior plaza. Nonetheless, the overall idea seems to be to design "a house in the forest".
Seung Isn't that the original reason for the existence of architecture? The architecture itself may be special, but the background function is more fundamental.

Pai "Where the architecture is revealed, where it is hidden," or "This area has no architecture, and that area does." Are such expressions meaningful during the actual design process?
Seung Most certainly. It is a necessary form of life created by the architect. There are cases in which the initial idea is carried on to completion, but more frequently, thoughts change countlessly during the design process.

Pai There seems to exist two polar architectural design processes. In one process, the idea of the project is first established and so- lidified, after which the design is developed accordingly. In the other process, various explorations and thoughts come together towards a later, final design. When I talk about Korean contemporary architecture, I state that Seung, H-Sang's process belongs to the former. Do you agree? How would you describe the process in your own words?
Seung It certainly fits with the former process. An architect must have core beliefs about architecture as well as our way of life. In the beginning of any project, I first establish the language of its fundamental concept. This is because the definition of this language will lead to the organization of thoughts and ideas that follow.

Pai The Shin DongYeop Literary Museum has a different type of composition. Furthermore, the circulation and the program differ greatly. Please explain this work in the context of these observations.
Seung The topic of Shin DongYeop Literary Museum had to be Shin DongYeop. Often, a literary museum shows that the archi-

不相同。请以这些视角为背景来解释这个工作。

　　承孝相：申东烨文学博物馆的主题必须是申东烨。通常情况下，一座文学馆表明建筑要优先于它的主题。这在姬路市文学博物馆有所体现，由安藤忠雄设计。首先，我需要研究申东烨，这有助于我了解他对我们土地的热爱。

　　现场的情况令人失望。我设想能有流动的BaekMa河和壮丽的、充满了历史遗迹的景色。但在现实中，它只是一个不起眼的小镇的一处平庸的场地。经过一番商议，我们决定为申东烨建造一个全新的地形。我们的主题是他所居住的不起眼的房子，与我们新建的景观形成了对比。其结果是，房子看起来比它的原来背景漂亮多了。而且，建筑在这里是隐藏的。

　　Pai：访客该如何围绕建筑移动呢？
　　承孝相：跟随展览馆展示的方向，参观者被引导到一个庭院。在这

申东烨文学博物馆
Shin DongYeop Literary Museum

里，顾客自然会遇见一条通往建筑物顶部的路径。一路上，游客可以欣赏屋顶的风景，而他们又会自然地被引领回地面。然后，他们会发现OkSang Lim安装的写满语言的旗子，这些旗子引导访客再次到达出发点。

　　Pai：在即将结束我们的谈话之前，我想请教一个一般性的问题。目前，Iroje建筑和规划事务所设计小项目，如HyunAm，同时也设计大型商业项目。另外，各个项目在韩国和中国同时开发。考虑到你的建筑思想大约每10年出现一个转折点，当前你面对的挑战是什么？

　　承孝相：目前还不是合适的时机来阐述。我们的工作要考虑公众利益和与社会的关系，这已成为一个重要的任务。如果我以前通过言语和意识形态提出过这些问题，那么那时就是行动日。我是一名建筑师，所以我设法通过建筑来找解决办法。

tecture was prioritized over its subject. This is exemplified in the Himeji City Museum of Literature designed by Tadao Ando. First, I needed to study Shin DongYeop. This led to an understanding that he loved our land.

The situation of the site was disappointing. I expected a flowing BaekMa River and magnificent scenery full of historical remains. But in reality, it was a banal fragment of a humble small town. After some deliberations, we decided to create a new topography for Shin DongYeop. Our subject was the humble house in which he resided, to which our new landscape was contrasted. As a result, the house looks much more beautiful than in it's original setting. Again, the existence of architecture here is hidden.

Pai How do visitors move around the building?
Seung Following the direction of the exhibition display, the visitor is led to a courtyard. Here, the visitor naturally meets a path to the top of the building. Along the way, visitors appreciate the scenery on the rooftop while they are naturally led to return to the ground. Then they meet the Flags of Language installed by OkSang Lim, through which the visitor reaches the starting point again.

Pai Finishing up our conversation, I'd like to ask a general question. Currently, Iroje designs small projects such as HyunAm as well as large-scale, commercial projects. Also, the various projects are in concurrent progress in both Korea and China. Considering that a turning point in your architectural thought has occurred approximately every 10 years, what are the current challenges that you face?
Seung It isn't yet the right moment to elaborate. It has become an important task in our work to consider the relation with public interest and with society. If I had previously raised such issues through words and ideology, these days are days of action. I am an architect, so I seek to find resolution through architecture.

Pai I know that they are incomplete, but are your comments applicable to the large-scale commercial projects in LA and SangAm-dong?

Pai：我知道这些建筑还未竣工，但您的宝贵意见适用于洛杉矶和SangAm-dong的大规模商业项目吗？

承孝相：当然。我们接受委托来设计某大型商业大楼，我们形成了一个计划。客户曾接受过其他的计划，并表示对我们的计划不感兴趣。他们建议我们重新匹配另一个计划。这些计划是基于图像的，和规模无关。当然，我们拒绝了建议。这是在委托设计方面错过的机会，也并非是有意针对我们的。

Pai：MoHeon的多层次，HyunAm扎根的大自然……，这些也都是规模的问题？

承孝相：在它们还是建筑理念之前，它们就是一个规模的问题。GunUi的辽阔大自然使层次划分是不必要的。总之，房屋的小规模根本不是问题。MoHeon和HyunAm处于完全不同的背景中，而我们的方法也相应地不同。此外，还有可能是有关功能的一个方面，但它绝对与景观相关联。

Pai：当然，MoHeon和HyunAm的确如此，但即使是对于BaekSa村和乐天购物中心来说，你仍然坚持"它们可以代表建筑"这个理念吗？

承孝相：我是一名建筑师。我相信我能解决我被世界给予的所有建筑问题，这是一个荒谬的想法吗？不过没关系的。

测试建筑的力量

"我相信我能解决我被世界给予的所有建筑问题"。这就是当我们结束谈话时，承孝相的说法。我想以下列方式来重申这句话，即"建筑存在于世界的所有问题中"。仅仅建筑是不足以解决问题的，但它肯定是必要的。事实上，我们需要建筑。承孝相是一位能够挖掘建筑潜力的建

Seung Certainly. We were commissioned a certain large commercial building and we created a plan. The client had previously received other plans, and expressed unfamiliarity with ours. They suggested that we reformat to match another plan. Those plans were image-based and unrelated with scale. Of course, we rejected the suggestion. It was a missed opportunity in terms of design commission, but it was not meant for us.

Pai The many layers of MoHeon, the great nature in which HyunAm is rooted... are these also matters of scale?

Seung Before they are architectural concepts, they are a matter of scale. The vast nature of GunUi made layer partitioning unnecessary. In conclusion, the small scale of the house was not problematic at all. MoHeon and HyunAm are in completely different settings, and our approaches differed accordingly. There also may have been an aspect concerning the program, but it was absolutely concerned the landscape.

Pai Certainly true to MoHeon and HyunAm, but even for BaekSa Village and Lotte Shopping Center, do you continue to hold the belief that "they may represent architecture?"

Seung I am an architect. I believe that I can solve all problems that I am given in this world through architecture. Is that an absurd thought? It does not matter.

Testing the Strength of Architecture

"I believe that I can solve all problems that I am given in this world through architecture." This was Seung, H-Sang's statement as we concluded our conversation. I would re-state this in the following manner: "architecture exists in all problems of the world." Architecture is not solely sufficient to solve problems, but it is certainly necessary. Indeed, we need architecture. Seung, H-Sang is an architect who explores this architectural potential. Of course, he studies that possibility through his particular method of architecture. As a result, many point out the many different kinds of architecture in this changing world and criticize Seung, H-Sang as an inflexible architect. I do believe that Seung, H-Sang's architecture is very "strong." But I do not believe that the exploration

筑师。当然，他是通过特定的建筑方式来研究这种可能性，其结果是，很多不同风格的建筑从不同的方面在改变这个世界和文明。并且我将其认定为一名固执的建筑师，我相信他的建筑是非常强大且坚固的。但我不相信对如此宏伟的建筑的探索的可能性已经达到了极限。相反，在韩国建筑的特定环境下，我认为这样的探索才仅仅进行了20年。作为一个例子，我想谈谈我和承孝相合作的经历。那是在2011年第四届光州双年展的安装设计阶段。承孝相和艾未未联合负责这个项目，我是负责协助这两名负责人来协调整个进程的总策展人。这是决定展览布局设计的重要时刻。在2011年1月，我们在艾未未的北京办公室里开了一次整体会议。在安装策略上，两名负责人和策展人之间发生了很多争论。当时，艾未未建议将这个展览建成一个统一的"超市"。相反，承孝相将双年展想象成带有不同结构和城市特色的空间。抚今追昔，艾未未要用一个中立的顺序来定位布展，而承孝相则希望设计一个视觉感受不到的平面顺序，一种通过空间形式来感受的顺序。换言之，承孝相希望通过建筑来解决安装问题。在许多策展人和艺术家的理念需要精细的协调的情况下，他的策略在一种条件下太难实施。我自己对他的策略持不确定的态度。然而，安装策略按照承孝相的方向进行了，主题和安装设施的协调简直就是痛苦的千夜噩梦。然而，由此产生的展览场地是一个给人惊喜、引导人们去发现的地方。艺术家和策展人的各种形式、对象、功能、文本、音乐、视频和表演都结合建筑秩序，创造一处城市般的空间。每个策展人的独特个性都融在一起，且和建筑的结合，创造了一处完全不同的展览空间。在这种充满了多样的主题和思维模式的情况下，强大的建筑的潜力得到了证实。我亲身经历了这一切，虽然很困难，但承孝相的建筑同时可以很灵活。

of the possibility of such strong architecture has reached a limit. Rather, in the particular context of Korean architecture I would argue that it has been barely 20 years since such exploration has begun. As an example, I'd like to talk about my own experience in collaborating with Seung, H-Sang. It was during the installation design phase of the 4th GwangJu Design Biennale in 2011. Seung, H-Sang and Ai WeiWei were co-directors and I was the chief curator responsible for assisting the two directors in coordinating the entire progress. It was an important moment in which the layout of the exhibit was to be decided. In January of 2011, we had an overall conference in Ai Weiwei's office in Beijing. There was much debate between the two directors and curators on the installation strategy. At the time, Ai WeiWei suggested that the exhibit be made into a homogenized "supermarket." In contrast, Seung, H-Sang imagined the Biennale as a space with heterogeneous architectural and urban characteristics. Thinking back, Ai WeiWei wanted to position the exhibit in a neutral order, whereas Seung, H-Sang desired a planar order undetectable to the eye, an order that would be felt through spatial form. In other words, Seung wished to resolve the installation through architecture. His strategy was too difficult in a situation in which the ideas of many curators and artists needed to be finely coordinated. I was myself uncertain about his strategy. Nevertheless, the installation strategy progressed in Seung, H-Sang's direction, and the coordination of the themes and installations were as painful as a thousand nights of nightmare. However, the resulting exhibit was a place of amazing discovery. The artists and curators' various forms, objects, programs, texts, music, videos, and performances were combined with an architectural order, creating a city-like space. The combination of each curator's unique personality, combined with architecture, had created a completely different exhibition space. In a situation full of diverse topics and mind-sets, the potential of a strong architecture was confirmed. I experienced first-hand that Seung, H-Sang's strong architecture, though difficult, can at the same time be flexible. *Hyungmin Pai*

承孝相 Seung, H-Sang

HyunAm, 一座黑色的小屋

依旧位于自然界中的房子

曾经，建筑师在大邱的SangYuk洞为一个想在新空间享受新生活的人设计了一座小房子和一座花园，分别命名为"MoHeon"和"SaYaWon"。后来，这个人开始在GunUi的一个1 000 000m²的山区修建一个野生植物园。他移植和砍伐树木了一段时间，同时观察自然的变化，并且想在那里修建一座小屋。场地坐落在山的中间，那里除了大山和天空其他什么都看不到。它面向西方，一定能给人们一个看到太阳落到水库之上的机会，并且在冬季沿着太阳的轨迹给人以温暖。

他希望他的植物园会成为一处适合沉思的地方。这就是为什么他建造了五个小型的、沿着穿过茂密冷杉林道路的人工花园。这些花园的规模非常小，但是这些花园的斜坡和平面间的和谐性会触发冥想。

在花园的末端，人们会看到一个考顿钢结构，带有一条长长的直线形斜坡。在这个结构的上层，人们会看到一个假山，而在下层，则会看到一个山谷的底部。人们不知道在门和入口处会看到什么，而且他们会发现自己身处雄伟的自然景观的深深环抱之中。那里没有建筑理念的存在，只有人们和自然处在完全的寂静中。

如果人们足够幸运，在日落时分，便会看到美丽的红色。当人们爬上山坡，坐在草丛中的考顿钢椅上，冷意袭来，便会觉得自己是自然的一部分。这是一个完全属于自己的沉思时刻。这就是为什么他称此为"HyunAm"，意味着黑色且阴暗的小屋。

当人们从远处看这座房屋时，便会发现它从地面伸出，就像它等待了很长时间从地下走出来，又进入世界。这就是为什么这座房子仍然保持不完整的原因。

HyunAm, A Black Cottage

The House That Remains Still in the Nature

Once the architect designed a small house named "MoHeon" and a beautiful garden named "SaYaWon" in SangYuk-dong, DaeGu for a person who wanted to enjoy a new life in a new space. Later, he began to build a wild botanical garden in a 0.3million-pyeong-wide mountain area in GunUi. He transplanted and cut trees for a long time, observing the change of the nature and wanted to build a small house to live there. The site is located in the center of the mountain where people can see nothing but the mountain and the sky. It faces the west, surely giving people a chance to see the sun going down over the reservoir and giving warmth in the winter season along the path of the sun.

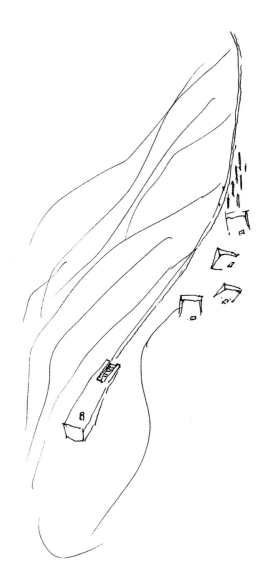

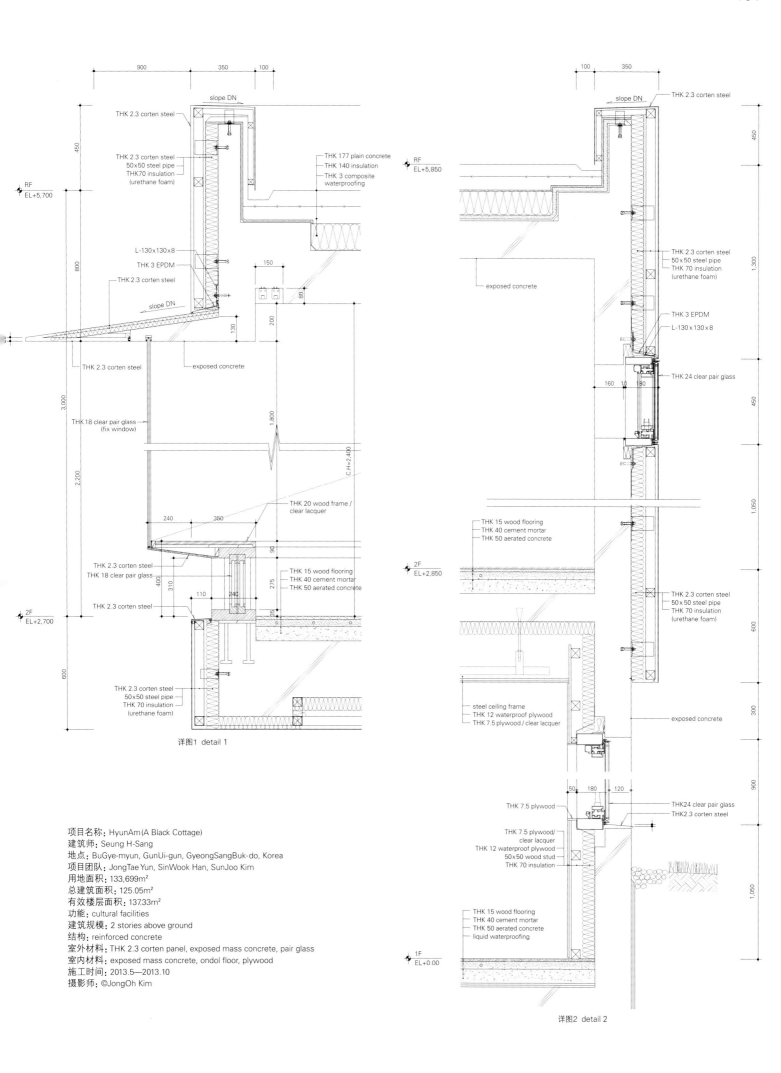

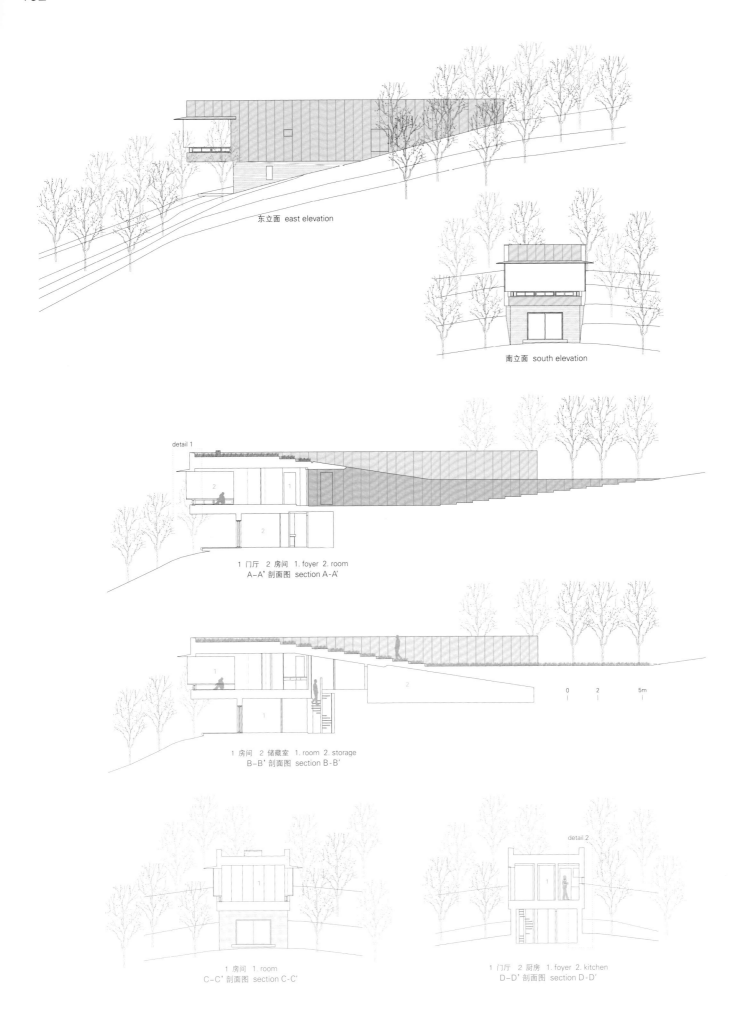

He hoped that his botanical garden would be a place for contemplation. This is why he made 5 small artificial gardens along the access road that passes through the dense fir tree forest. These are all small but the harmony between the slopes and planes of these gardens will be a trigger for meditation.

At the end of the gardens, people will see a corten steel structure with a long straight slope. On the upper part of this structure, people will see an artificial hill, while in the lower part, they will see the bottom of a valley. People have no idea on what they will see over the gate and enter there, and then they will find themselves in a deep embrace of the majestic natural landscape. No architectural concept exists there, only people and the nature are in complete silence.

If people are fortunately there when at sunset, they will see the beautiful red color of it. When they climb on the hill and sit down on a cold corten chair in the grass forest, they will feel like they are a part of the nature. This is a completely solitary moment, the moment for meditation. This is why he named it "HyunAm" which means a black and shady cottage.

When people see this house at a distance, they will find it protruded from the ground as if it has just come out of the underground into the world after waiting a long time. That is why this house remains still with incompleteness. Seung, H-Sang

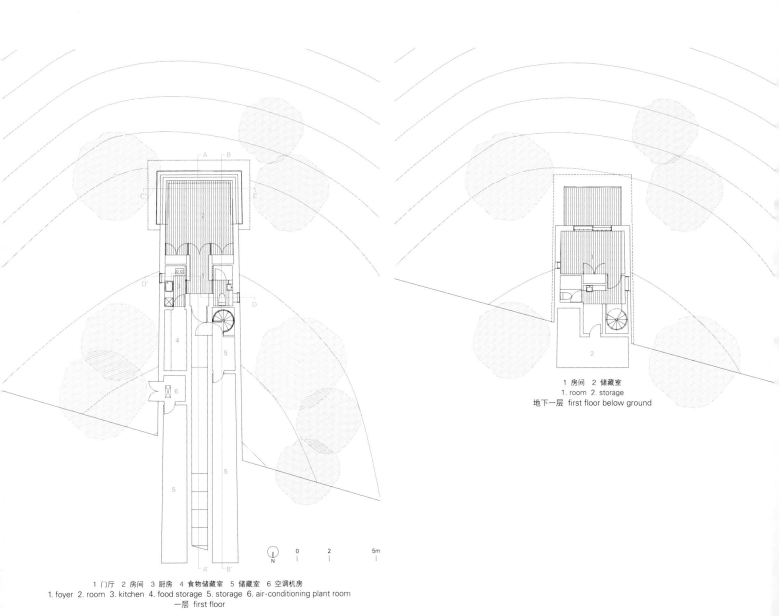

1 门厅　2 房间　3 厨房　4 食物储藏室　5 储藏室　6 空调机房
1. foyer 2. room 3. kitchen 4. food storage 5. storage 6. air-conditioning plant room
一层 first floor

1 房间　2 储藏室
1. room 2. storage
地下一层 first floor below ground

承孝相 Seung, H-Sang

申东烨文学博物馆

不纪念建筑本身

"是谁说,我看到了天空,是谁说,我看到了天空,那是晴朗的,没有云彩……

纪念写这首诗的申东烨(1930—1969)的文学博物馆就建在位于他的家乡韩国扶馀的房子旁边,他在那里生活了一辈子。他的住处曾经是一座三室的茅草屋顶房子,而现在是一座破旧石板屋顶覆盖的房子,似乎表明了在那个不和谐的时代,诗人生活中的悲伤情绪。但是,纪念那位抱怨着时代的不如意、大喊着"走开,你这空虚的躯壳!"的诗人不是建造这座纪念馆的唯一目的,我觉得这座博物馆给了我们一个机会,以纪念自己和我们所处的在这片土地。这所房子可以指引我们跟随它的通道回到我们最初开始的地方,从而给我们一个想起他和再次找到自己的机会。

在这种意义上来讲,这座博物馆的通道设置成一个圆形结构。在你穿过花园进入博物馆,看到他的作品时,你已经在庭院里了。当你爬上那里的楼梯,你会看到一片新的土地,一直通往上层,在那里四处走走,你会走到低处,再次看到一个飘满旗帜的广场,旗帜上都是诗人的诗歌。你会被这些词的美感所吸引,然后发现自己已经回到了出发的地方。

这座建筑仅仅是个媒介,它给了我们一个机会去体验这种经历,从这层意义上说,这在任何建筑理念中都不存在,这座混凝土结构的建筑物仅仅是一个实体背景,并不是纪念建筑本身。

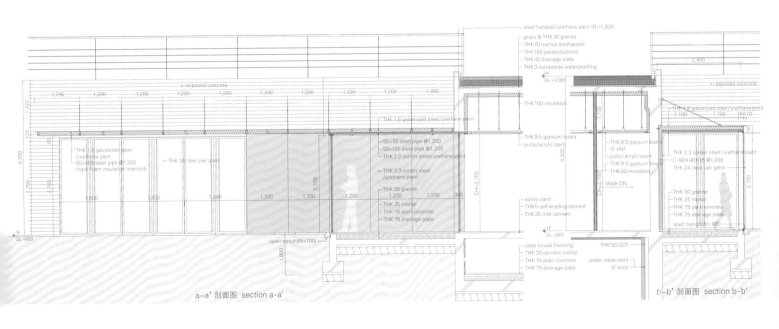

a-a' 剖面图 section a-a'   b-b' 剖面图 section b-b'

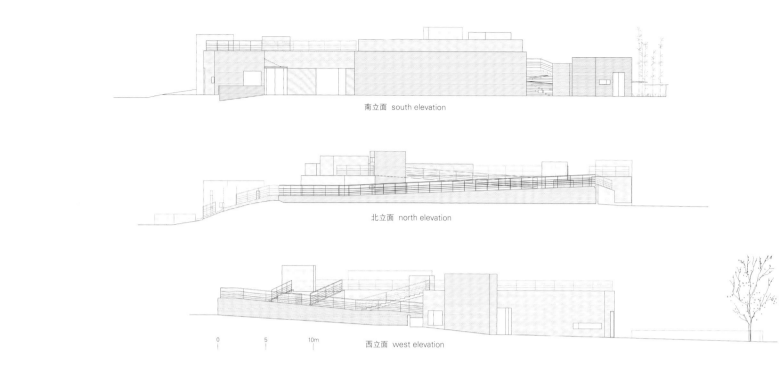

南立面 south elevation

北立面 north elevation

西立面 west elevation

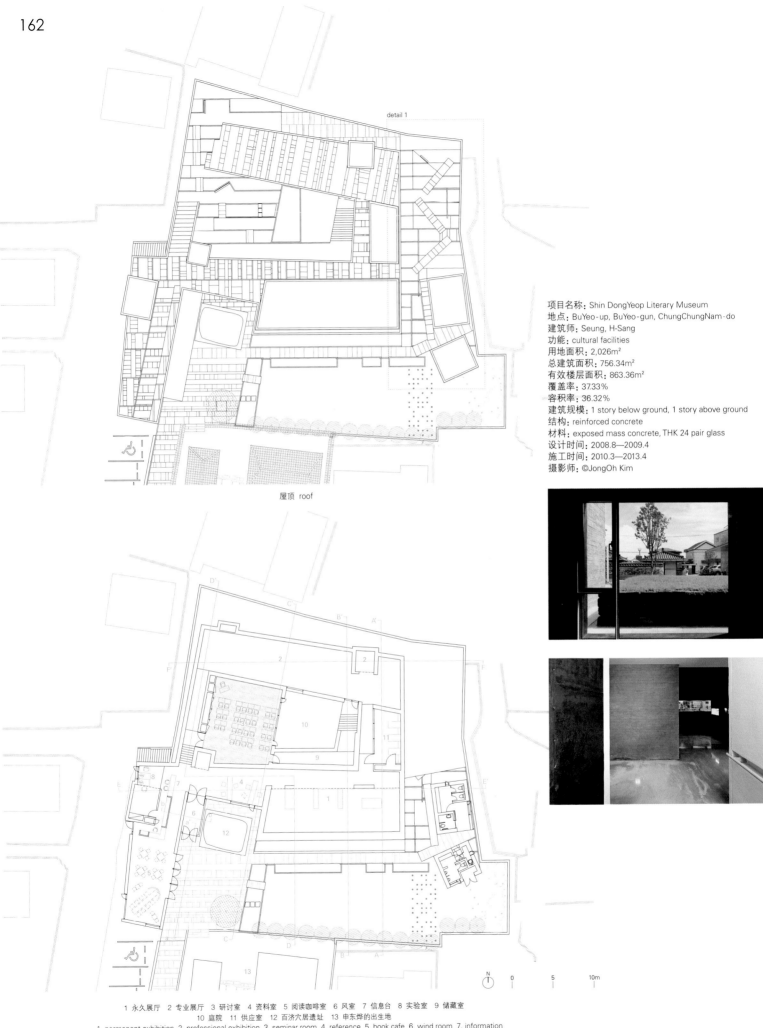

屋顶 roof

1 永久展厅 2 专业展厅 3 研讨室 4 资料室 5 阅读咖啡室 6 风室 7 信息台 8 实验室 9 储藏室
10 庭院 11 供应室 12 百济穴居遗址 13 申东烨的出生地

1. permanent exhibition 2. professional exhibition 3. seminar room 4. reference 5. book cafe 6. wind room 7. information
8. laboratory 9. storage 10. courtyard 11. supply room 12. BaekJe pit dwelling ruin 13. Shin DongYeop's birthplace

一层 first floor

项目名称：Shin DongYeop Literary Museum
地点：BuYeo-up, BuYeo-gun, ChungChungNam-do
建筑师：Seung, H-Sang
功能：cultural facilities
用地面积：2,026m²
总建筑面积：756.34m²
有效楼层面积：863.36m²
覆盖率：37.33%
容积率：36.32%
建筑规模：1 story below ground, 1 story above ground
结构：reinforced concrete
材料：exposed mass concrete, THK 24 pair glass
设计时间：2008.8—2009.4
施工时间：2010.3—2013.4
摄影师：©JongOh Kim

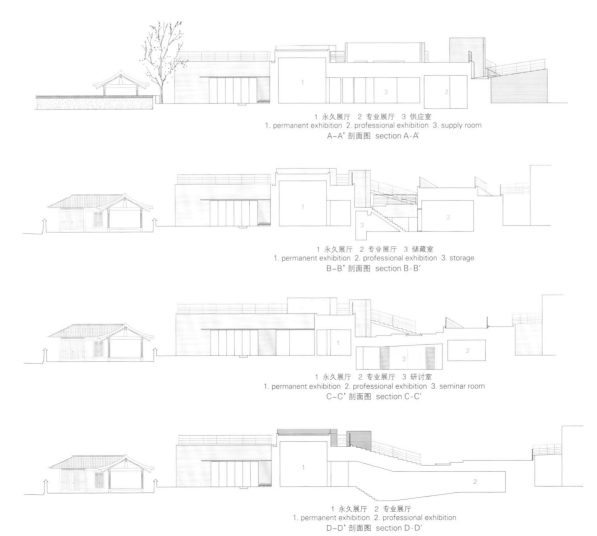

1 永久展厅 2 专业展厅 3 供应室
1. permanent exhibition 2. professional exhibition 3. supply room
A-A' 剖面图 section A-A'

1 永久展厅 2 专业展厅 3 储藏室
1. permanent exhibition 2. professional exhibition 3. storage
B-B' 剖面图 section B-B'

1 永久展厅 2 专业展厅 3 研讨室
1. permanent exhibition 2. professional exhibition 3. seminar room
C-C' 剖面图 section C-C'

1 永久展厅 2 专业展厅
1. permanent exhibition 2. professional exhibition
D-D' 剖面图 section D-D'

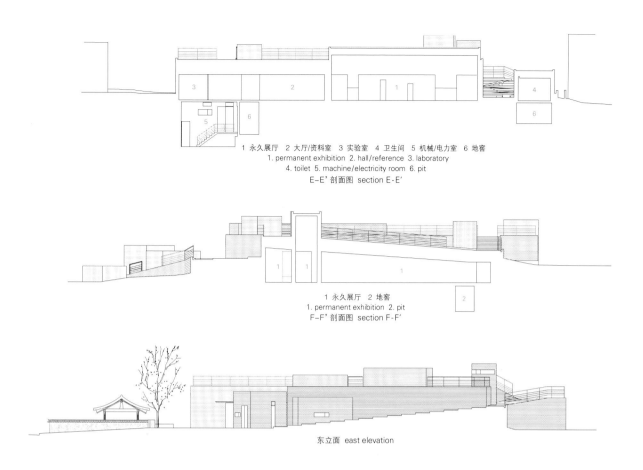

1 永久展厅 2 大厅/资料室 3 实验室 4 卫生间 5 机械/电力室 6 地窖
1. permanent exhibition 2. hall/reference 3. laboratory
4. toilet 5. machine/electricity room 6. pit
E-E' 剖面图 section E-E'

1 永久展厅 2 地窖
1. permanent exhibition 2. pit
F-F' 剖面图 section F-F'

东立面 east elevation

Shin DongYeop Literary Museum

The House That Does Not Commemorate Itself

"Who is it that says, 'I saw the sky'. / Who is it that says, 'I saw the sky / that was clear with no cloud....'"

The literary museum that commemorates Shin DongYeop (1930~1969) who wrote this poem was built right next to the house located in his hometown BuYeo, Korea where he lived in his lifetime. It was once a thatched-roof house with 3 rooms and is now a shabby slate-roofed house which seems to show the sadness of the poet who lived his life in the era of disharmony. However, it is not the only purpose of this memorial museum to commemorate the poet who cried out the era of unhappiness, saying *"Go away! You, empty shell!"* I think this museum gives us a chance to commemorate ourselves and the ground where we are now standing. This house makes us follow along the passage of it and then return to the place where we started and thereby gives us a chance to remind him and find ourselves again.

In this sense, this passage of this museum is a circular structure. After you enter the museum through the garden and see his works, you will be in the courtyard. When you goes up the stairs there, you will see a new land that will lead to the upper level, making you walk around there and then lead to the lower level where you will see a square that is full of flags flying with the words of his poems. You will be fascinated by the beauty of these words and then found finally that you are in the place where you started.

This architecture is merely a medium that gives us a chance to have this kind of experience and thus it does not exist in terms of any architectural concept. In this sense, this concrete-structured building is merely a physical background, not commemorating itself. Seung, H-Sang

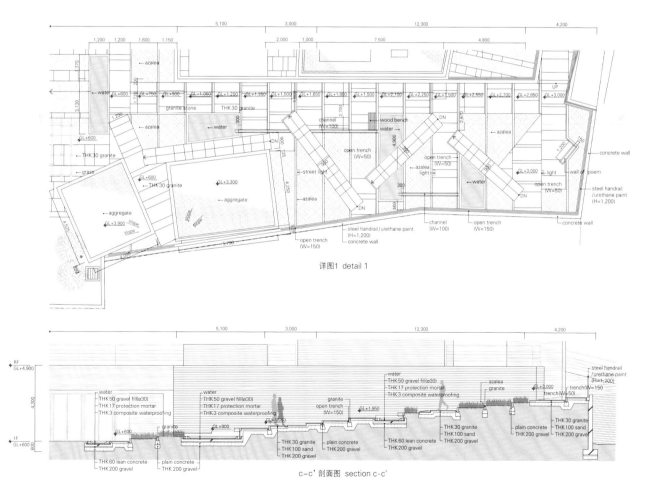

详图1 detail 1

c-c' 剖面图 section c-c'

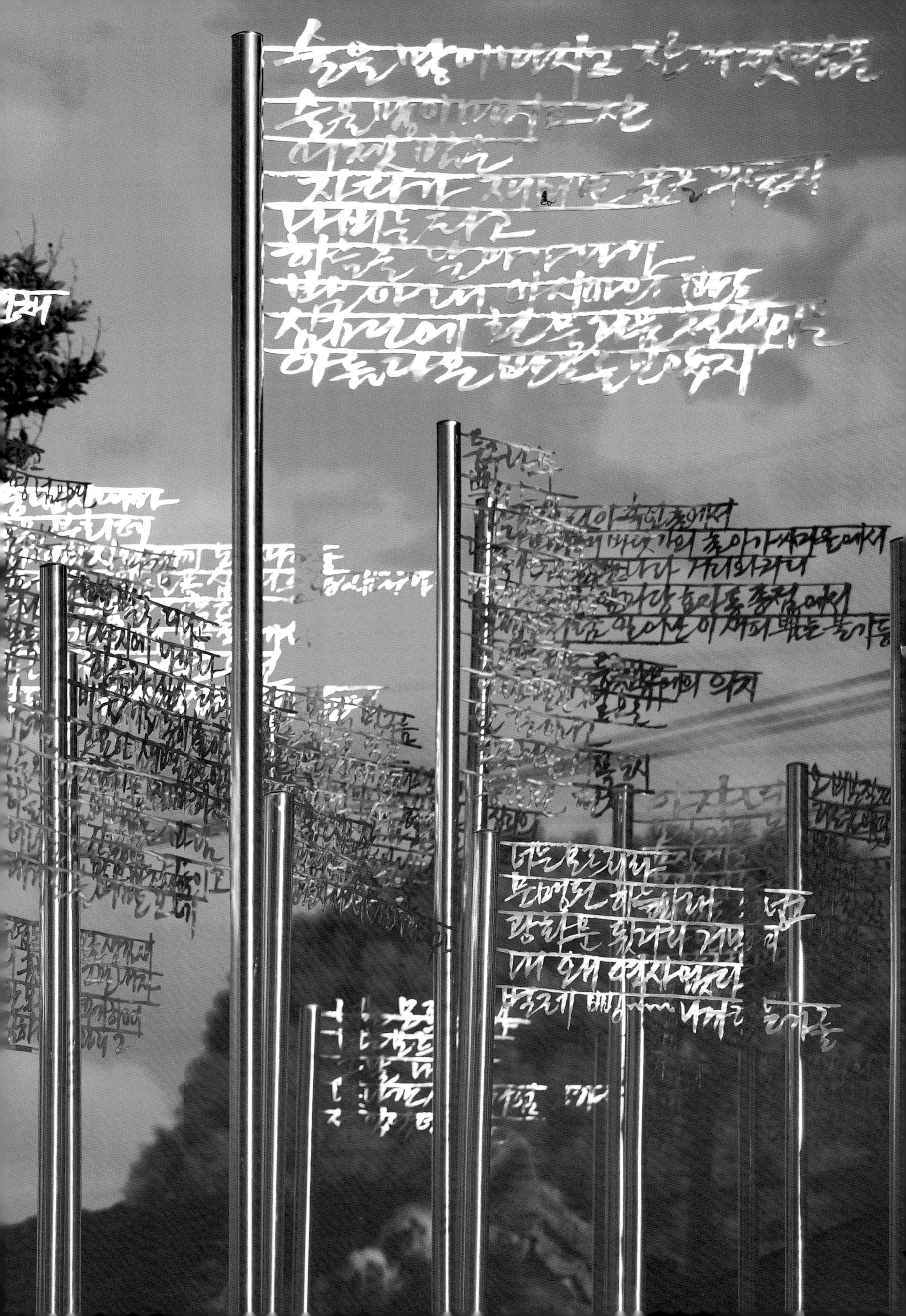

平度市住宅文化馆

不存在的建筑

平度市是一个拥有3000年历史和150万人口的小城市。位于青岛北郊,这座城市最近经历了快速的城市化进程。政府机关从市区转移到新城区,已经启动发展主要老街道的总体规划。

人们期望在这座古老的、拥有几千年历史的城市里发现古代的遗迹,不幸的是,他们意识到,建设在那里的公寓楼和政府机关建筑已经毁了所有的遗迹。然而,幸运的是,古代地图显示,很多旧的道路仍在使用。因此,基于这个框架,建筑师建立了城市修复的总体规划,重在"读风景",并命名为平度市历史区重建计划。

负责重建计划的万科集团被要求重新设计第一座建筑,把它作为一个居住文化中心,用于项目的公共关系建设。施工场地是一个附属于市政府的一个小型建筑场所的一部分,周围是茂密的树林,长满了2~3年的小树,这些都要求必须保留。所以建筑师没有选择,只能在一片空旷的土地上建造新建筑。但问题是,按照要求,这座新建筑要连接道路,让市民可以认出它。为此,混凝土框架安装在道路上的边界线上。这些混凝土框架被赋予一个功能:作为街头画廊,并且通过建在入口通道上的画廊来与建筑相连。然而,所有这些结构仅仅是一个工具,用来界定建筑物周围的风景优美的树林的空间界限,但它们在建筑理念方面并不存在。在大楼里,游客可以看到一间虚拟样板房,所在的空间轴有意地把它与引起游客更多兴趣的真正的内部空间分隔开来。不过,最重要的因素是围绕在那些告诉我们过去故事的建筑周围的树林。因此,这必须被视为在未来所有的建筑工作中最重要的原则。

Pingdu Housing Culture Center

The Building That Does Not Exist

Pingdu is a small city with 1.5 million populations and of 3000 years' history. Located in the northern outskirts of Qingdao, this city has recently experienced rapid urbanization. The government offices moved from the downtown area to the new town area and their master plan to develop the old main streets was launched.

项目名称:Pingdu Housing Culture Center
地点:Pingdu, Qingdao, China
建筑师:Seung, H-Sang
项目团队:SungHee Kim, KiTae Lee, MoonHo Lee, TaeYong Kim
功能:cultural facilities
有效楼层面积:2,340.8m²
建筑规模:2 stories above ground
结构:reinforced concrete
高度:9.6m
室外材料:THK 24 pair glasses, THK 2.3 corten panel, basalt
室内材料:acrylic paints, stone, wood
设计时间:2011.4—2012.1
施工时间:2012.10—2013.7
摄影师:©JongOh Kim

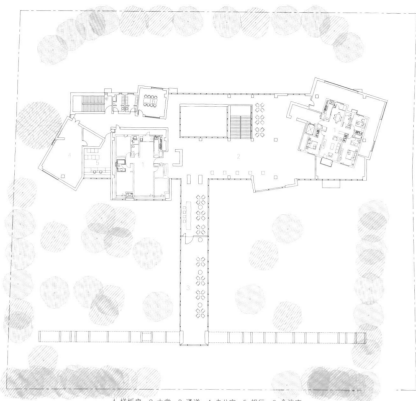

1 样板房 2 大堂 3 通道 4 办公室 5 银行 6 会议室
1. model house 2. lobby 3. aisle 4. office 5. bank 6. conference room
二层 second floor

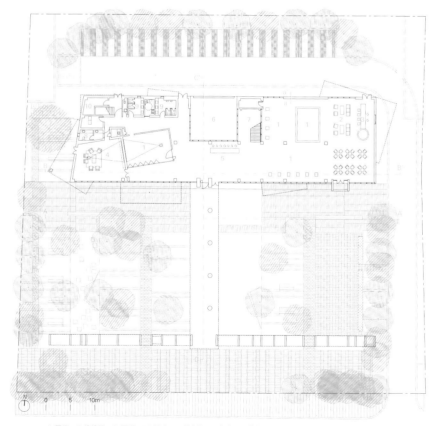

1 展厅 2 咨询区 3 酒吧 4 VIP室 5 信息台 6 庭院 7 楼梯 8 VIP卫生间 9 卫生间 10 储藏室
1. exhibition 2. consultation area 3. bar 4. VIP room 5. information 6. courtyard 7. staircase 8. VIP toilet 9. toilet 10. storage
一层 first floor

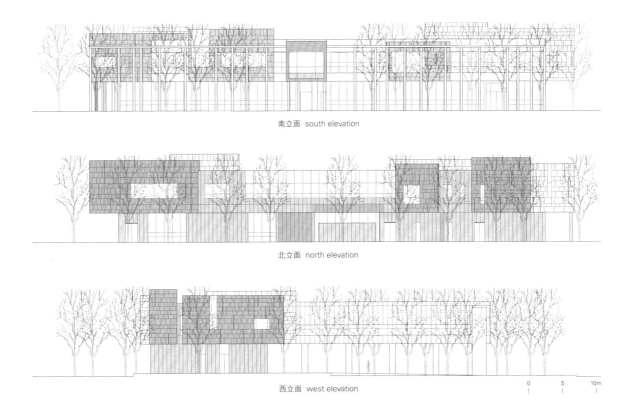

南立面 south elevation

北立面 north elevation

西立面 west elevation

People expected that they would find relics of ancient times in this old city of several thousand years' history and unfortunately, they realized that the apartment houses and government offices built there had already ruined all of the relics. However, a fortunate thing is that ancient maps show that many of the old roads are still used. So based on this framework, a master plan for urban restoration was established focusing on "landscript" and named Pingdu Historical Area Regeneration Plan.

Vanke Group that took charge of the regeneration plan was requested to redesign the first building to use it as a housing culture center for public relations of the project. The construction site was a part of the site for a small building attached to the city hall with dense woods that were all over 2~3 years old and required to be preserved. So there was no option but to build the new building on an empty land. The problem was that it was required to connect this new building to the road so that citizens can recognize it. For this, concrete frames were installed on the boundary line of the road. These concrete frames were given a function as a street gallery and connected with the building through the gallery built over the access road. However, all these structures are merely a tool that defines the spatial boundaries of the beautiful woods around the building, not existing in terms of architectural concept. In the building, visitors can see a virtual model house where the spatial axis is misaligned intentionally to separate it from the real space which arouse a little bit more interest in the inside area. Nevertheless, the most overwhelming factor is the woods around the building that tells us about the past. Therefore, this must be regarded as the most important principle of all architectural works in the future. Seung, H-Sang

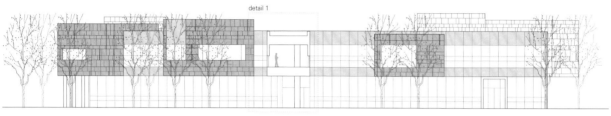

A-A' 剖面图　section A-A'

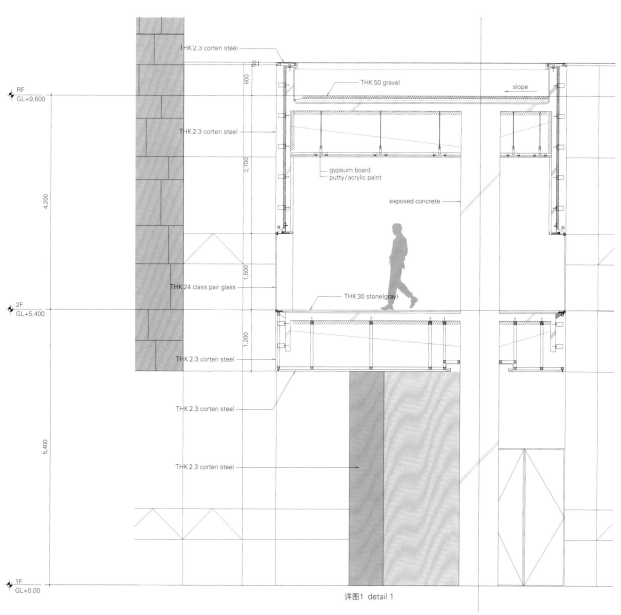

详图1 detail 1

1 VIP室 2 信息台 3 楼梯 4 展厅 5 酒吧 6 办公室 7 银行 8 样板房 9 大厅
1. VIP room 2. information 3. staircase 4. exhibition room 5. bar 6. office 7. bank 8. model house 9. lobby
B-B' 剖面图 section B-B'

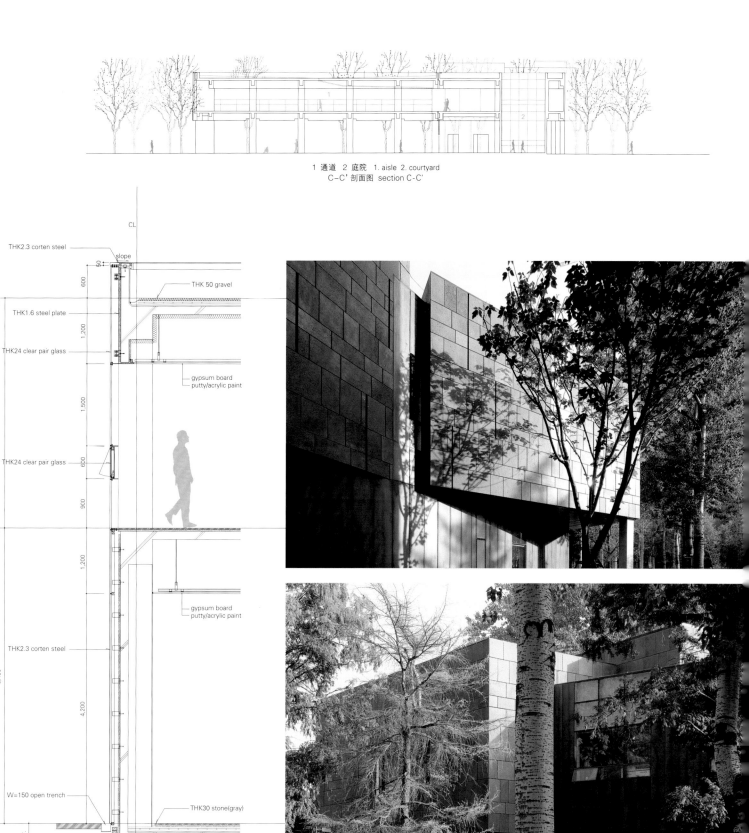

1 通道 2 庭院 1. aisle 2. courtyard
C-C' 剖面图 section C-C'

a-a' 剖面图 section a-a'

>46
A4 estudio
Was founded by Leonardo Codina[left] and Juan Manuel Filice[right] in 2004. Leonardo Codina received a M.Arch from Pontifical Catholic University of Chile(PUC) in 2005. Juan Manuel Filice also received a M.Arch from the University of Navarra in Spain in 2003. They have been teaching at the PUC since 2006. They pursue an architecture that is understood from the beginning as a platform for personal growth, professional and academic, which considers each request for proposal as an opportunity, contribution and research.

>>106
Laurens & Loustau Architectes
Is a French architecture firm, based in Toulouse, co-founded by Marc Laurens & Pierre Loustau in 1996. Marc and Pieerre both graduated from Toulouse National School of Architecture together in 1990 and have been teaching at the university since 2008. The work of the agency has been awarded in the category of Midi-Pyrénées Architecture prize for the several school projects such as Montastruc High School in 2003, Pole Graphic Arts High School Jolimont in 2005 and the Gymnasium INPT in 2009.

>>54
noname29
Alfredo Payá Benedito was born in Alicante, Spain and studied at the Technical Superior School of Architecture of Madrid. Has worked as projects professor in several educational institutions in Alicante, Catalonia and Madrid. Has been chosen for the Venice Biennale, in 2000 and 2002, representing the Spanish Pavilion. At the beginning of 2006, he founded noname29 where he develops his work at present. Is a winner of different national and international competitions. His works have been awarded in many occasions and reviewed in different publications and national and international exhibitions.

>>76
Abalo Alonso Arquitectos
Elizabeth Abalo[left] and Gonzalo Alonso[right] received a M.Arch from the University of Navarra, Spain. Gonzalo Alonso has been an Associate Professor at Cesuga, University College Dublin from 2008 to 2011. Both of them taught at the University of A Coruña from 2011 to 2012. They have been working together since 1997 and obtained several awards and distinctions; the Coag Award 2012 for Rubido Romero Foundation, the Aplus Award 2011 for the Oleiros Health Center, the Spanish National Heritage Prize 2009 for the San Clemente Square and the Galician Rodriguez Peña Award 2007 for the San Pedro 78 House.

>>84
JKMM Architects
Was established in 1998 by four partners; Juha Mäki-Jyllilä, Samuli Miettinen, Asmo Jaaksi, Teemu Kurkela from the left in Finland. Tries to make architecture with exceptional architectural and technical quality and to bring together innovation, intelligence and common sense. Operates actively in various areas and scales of architecture designing buildings, interiors, furniture, urban environments as well as renovations. Has received over 56 prizes in architectural competitions in Finland. Believes that buildings should be at their best even after decades of use and sustainable development is the key responsibility. Is developing a tailor-made solution for every project. Uses no standard methods, but create new processes.

>>128
Guzmán de Yarza Blache
Guzmán de Yarza Blache holds a Bachelor Degree in Architecture from the University of Navarra in 2002. Founded a Madrid-based office, J1 Architects with three other partners. Currently undertakes projects and supervises construction works in Spain. As a visual artist as well as architect, he develops a career and a combination that enriches both practices and allows unexpected cross relations. Currently he teaches at the USJ School of Architecture in Zaragoza and also at the IE School of Architecture and Design in Segovia and Madrid.

Hyungmin Pai
Graduated from the Department of Architecture, and the Graduate School of Environmental Studies, Seoul National University. He received his Ph.D from the History, Theory, and Criticism program at MIT. Twice a Fulbright Scholar, he is professor at the University of Seoul and Chair of the Mokchon Architecture Archive. He was visiting scholar at MIT and London Metropolitan University and has taught and lectured internationally. His books include The Portfolio and the Diagram(MIT Press), Sensuous Plan: The Architecture of Seung, H-Sang (Dongnyuk), and Key Concepts of Korean Architecture (Dongnyuk). For the Venice Biennale, he was twice curator for the Korean Pavilion (2008, 2014) and a participant in the "Common Pavilion" project. He was guest curator at Aedes Gallery (Berlin) and the Tophane Amire Gallery (Istanbul) and Head Curator for the 4th GwangJu Design Biennale.

©Kim SukGoo

>>136
Seung, H-Sang
Graduated with his master's degree from Seoul National University and studied at Vienna University of Technology. After working for SwooGeun Kim from 1974 to 1989, established Iroje Architects & Planners in 1989. In 2002, was the first architect to be named Artist of the year by the National museum of Contemporary Art, Korea, where he held a grand private exhibition. Was a visiting professor of North London University and taught at Seoul National University and at Korea National University of Art. In 2007, Korean government honored him with 'Korea Award for Art and Culture', and was commissioned as director for GwangJu Design Biennale 2011 after for Korean Pavilion of Venice Biennale 2008. Currently has been invited to Venice Biennale 2012.

Douglas Murphy
Studied architecture at the Glasgow School of Art and the Royal College of Art, completing his studies in 2008. As a critic and historian, he is the author of The Architecture of Failure (Zero Books, 2009), on the legacy of 19th century iron and glass architecture, and the forthcoming Last Futures (Verso, 2015), on dreams of technology and nature in the 1960s and 70s. Is also an architecture correspondent for Icon Magazine, and writes regularly for a wide range of publications on architecture and culture.

Aldo Vanini
Practices in the fields of architecture and planning. Had many of his works published in various qualified international magazines. Is a member of regional and local government boards, involved in architectural and planning researches. One of his most important research interests is the conversion of abandoned mining sites in Sardini.

>>36
Pereda Pérez Arquitectos
Was founded by Carlos Pereda Iglesias[left] and Oscar Perez Silanes[right] in Pamplona, where they conduct their business engaged in the research and realization of architectural projects. They have presented their works in conferences of different Spanish institutions and overseas. Their works have been published in various journals and national publications. They currently develop their educational works as associated lecturers of architectural projects at Zaragoza University.